T0345137

Continuous Monitoring for Hazardous Material Releases

This book is one in a series of process safety guideline and concept books published by the Center for Chemical Process Safety (CCPS). Please go to www.wiley.com/go/ccps for a full list of titles in this series.

Continuous Monitoring for Hazardous Material Release

Center for Chemical Process Safety
New York, New York

American Industrial Hygiene Association
Fairfax, Virginia

A JOHN WILEY & SONS, INC., PUBLICATION

It is sincerely hoped that the information presented in this document will lead to an even more impressive safety record for the entire industry. However, neither the American Institute of Chemical Engineers, its consultants, CCPS Technical Steering Committee members, their employers, their employers' officers and directors, nor the American Industrial Hygiene Association and its employees warrant or represent, expressly or by implication, the correctness or accuracy of the content of the information presented in this document. As between (1) American Institute of Chemical Engineers, its consultants, CCPS Technical Steering Committee and Subcommittee members, their employers, their employers' officers and directors, and the American Industrial Hygiene Association and its employees and (2) the user of this document, the user accepts any legal liability or responsibility whatsoever for the consequence of its use or misuse.

A Joint Publication of the Center for Chemical Process Safety of the American Institute of Chemical Engineers, the American Industrial Hygiene Association, and John Wiley & Sons, Inc.

Published by John Wiley & Sons, Inc., Hoboken, New Jersey
Published simultaneously in Canada

For general information on our other products and services or for technical support, please contact our Customer Care Department within the United States at (800) 762-2974, outside the United States at (317) 572-3993 or fax (317) 572-4002.

Wiley also publishes its books in a variety of electronic formats. Some content that appears in print may not be available in electronic formats. For more information about Wiley and AIChE products, visit our web site at www.wiley.com.

Library of Congress Cataloging-in-Publication Data:

Continuous monitoring for hazardous material releases / Center for Chemical Process Safety [and] American Industrial Hygiene Association.
    p. cm.
Includes bibliographical references and index.
ISBN 978-0-470-14890-7 (cloth)
1. Continuous emission monitoring. I. American Institute of Chemical Engineers. Center for Chemical Process Safety. II. American Industrial Hygiene Association.
TD890.C542 2009
660'.2804—dc22   2009001780

Printed in the United States of America

10 9 8 7 6 5 4 3 2 1

Table of Contents

Acknowledgments

The Center for Chemical Process Safety (CCPS) and the American Industrial Hygiene Association (AIHA) thank all of the members of the Continuous Monitoring Subcommittee for providing technical writing, technical input, reviews, guidance and encouragement to the project team throughout the preparation of this book. CCPS and AIHA also express appreciation to the members of the CCPS Technical Steering Committee and the AIHA Engineering Committee for their advice and support.

The CCPS staff liaison for this project was John Davenport. The subcommittee had the following members, whose significant efforts and contributions are gratefully acknowledged:

Chandran Achutan NIOSH	Eugene Lee US EPA/OEM
Sandra Barnes Chevron Texaco	Leo Old Smith Seckman Reid
Srikar Chunduri Advanced Micro Devices	Dennis Patrick Airgas
John A. Davenport CCPS/AIChE	Aimee Petrognani American Industrial Hygiene Association
Stephen W. Haines RRS Engineering	Tim Ross Intel
Joselito Ignacio US Coast Guard	Scott Swanson Intel
Veronica Kero Omega Environmental	Ronald Taylor Aramco
Steven E. Lacey University of Illinois at Chicago	Anthony Taylor Monsanto Company

The book Continuous Monitoring for Hazardous Material Releases was composed by the publication staff of AIHA, led by David Binswanger.

CCPS and AIHA also gratefully acknowledge the valuable suggestions and feedback submitted by the following persons who provided peer review comments on interim and final draft manuscripts:

Stanley S. Grossell, Process Safety and Design, Inc., Patrick Hogan, Honeywell Analytics, and Michael D. Scott, AE Solutions.

The constructive feedback provided by Brian D. Kelly, CCPS Emeritus Member, and detailed editing provided by Stephen W. Haines are especially appreciated.

Preface

The American Institute of Chemical Engineers (AIChE) has been closely involved with process safety and loss control issues in the chemical and allied industries for more than five decades. Through its strong ties with process designers, constructors, operators, safety professionals, and members of academia, AIChE has enhanced communications and fostered continuous improvement of the industry's high safety standards. AIChE publications and symposia have become information resources for those devoted to process safety and environmental protection.

AIChE created the Center for Chemical Process Safety (CCPS) in 1985 after the chemical disasters in Mexico City, Mexico, and Bhopal, India. CCPS is chartered with developing and disseminating technical information for use in the prevention of major chemical accidents. The center is supported by more than 125 sponsoring companies within the chemical, petroleum, and related process industries who provide the necessary funding and professional guidance to its technical committees. The major product of CCPS activities has been a series of Guideline Books and Concept Books to assist those implementing various elements of a process safety and risk management system. This book is part of the Concept Book series.

Founded in 1939, the American Industrial Hygiene Association is one of the largest international associations serving the needs of occupational and environmental health professionals practicing industrial hygiene in industry, government, labor, academic institutions, and independent organizations. AIHA members are devoted to achieving and maintaining the highest professional standards; work in conjunction with the American Board of Industrial Hygiene to promote certification of industrial hygienists; administer comprehensive education programs that keep occupational and environmental health and safety professionals current in the field of industrial hygiene; and operate several highly-recognized laboratory accreditation programs, based on the highest international standards. These programs help ensure the quality of the data used in making critical worker protection decisions.

The chemical industry commonly utilizes a number of systems which analyze ambient air for content of hazardous materials. A variety of equipment is available to perform this kind of monitoring, utilizing a number of different technical approaches. While much has been written about monitoring for occupational exposure, literature related to rapid detection of leaks and catastrophic releases is relatively scant, and the strengths and weaknesses of each are not often presented in vendor literature. Likewise the rationale for placement of area detectors is frequently poorly defined.

As a result, instances exist where monitoring equipment to detect catastrophic release is not been used appropriately, or the information generated by the system falls short of what it is capable of delivering. CCPS and AIHA therefore agreed to collaborate on this subject of mutual interest. The result is this practical guide to help the user select, implement, and get the most out of their catastrophic release detection systems.

1
Introduction

1.1 PURPOSE

The primary driver behind the development of early gas detectors was the mining industry, which was experiencing notoriously high mortality rates related to the presence of oxygen deficient, explosive and toxic atmospheres. In response to this problem, the first gas detectors were developed in the mid-1920.

Since then, the use of gas detectors has extended to other industries, technologies have advanced and our knowledge concerning the behavior of released vapors and gases has expanded. However, older gas detection systems that are not as reliable as the newer ones are still in use.

This guideline was developed with the intention of helping today's facility to get the most out of their gas detection systems. It includes guidance on:

- Available gas sensors, including basic descriptions of how each type works, their capabilities and their limitations.
- Establishing objectives for the use of gas detection systems and incorporating them into the facility's operating and emergency procedures.
- Placing gas detectors to maximize the potential for detection success during a release.
- Managing gas detection system design, alarm, and operating parameters.

1.2 SCOPE

This text is intended to assist the user to detect and respond to accidental release of hazardous materials. It is not intended to address:

- Integrated exposure sampling to assess employee exposure against documented standards i.e. 15 min STEL (Short Term Exposure Limit), or 8 hour threshold Limit Value (TLV) time weighted average etc.
- Background or continuous monitoring required for environmental regulatory purposes.

1.3 WHO WILL BENEFIT FROM THIS GUIDELINE?

Because gas detection may be utilized to meet a wide range of objectives, this Guideline will benefit many different people within an organization.

1. Corporate Leadership – Senior executives define the basis for the development of personnel and asset protection philosophies. Their commitment and recognition of the value of gas detection is vital to the implementation personnel and asset protection strategies.
2. Site Managers – Site managers are responsible for developing and maintaining the facility's gas detection philosophy and strategies. They are also responsible for developing emergency response and site evaluation policies.
3. Line Management – Line managers are responsible for maintaining gas detection systems and for ensuring personnel are trained on their use and limitations. They ensure that policies and procedures, including gas detection, are integrated and implemented. This includes testing and maintenance of the gas detection systems.
4. Project Managers – Project managers are responsible for executing projects, usually from design through startup and commissioning. A Project manager is responsible for determining the basic system design concepts to apply in the execution of a project. The Project manager is responsible for implementing the decisions and abiding by the project procedures associated with amending and adding to the gas detection system.
5. Engineers – Engineers are responsible for specifying and designing gas detection systems that meet their company's personnel and asset protection requirements. There is a lot of room for decision making when designing gas detection systems, making knowledge about their capabilities, limitations and modern design practices critical.
6. HSE Professionals – Health, safety, and environmental (HSE) professionals provide technical guidance to engineers and typically are in an assurance role for gas detection systems.
7. Emergency Response Personnel – Emergency response personnel provide guidance on how gas detection system should be integrated in the overall emergency response protocols.

2
Management

2.1 MANAGEMENT OVERVIEW

Many companies have gas detection systems, but very few actually manage their use. As a result, many have been plagued by:

- Inconsistent protection levels
- Inappropriate or over usage.
- Excessive installation and maintenance costs.
- Systems that could actually pose a hazard, as a result of:
  - Employees placing too much or too little confidence in them
  - Employees reacting to alarm activations inappropriately
  - Controls/safety devices reacting to alarm activations inappropriately

All of these situations can be avoided if a holistic approach to gas detection management is taken. This rarely requires ongoing full-time attention, but does require some upfront work to establish the management program, along with some continuous improvement efforts to keep it up to date.

When developing a gas and vapor detection management program, one must first answer four basic questions:

- Why do we use gas detection?
- What do we want to detect?
- What actions do we expect to be undertaken when a release is detected?
- How much should we spend on gas detection?

If the implementation team can answer these questions, put these answers into writing in the form of a procedure or standard, keep it up to date and stick to it, a lot of angst and confusion can be avoided and a greater return will be seen with respect to degree of protection and expenditures on gas detection.

The answers to these questions are discussed below, while the technical background for developing the answers can be obtained from Chapters 3 and onward.

2.2 WHY DO WE USE GAS DETECTORS?

Gas detection systems are typically utilized for one or more reasons including:

- Personnel Safety – These systems typically go beyond those required by regulatory agencies and are intended to assure that personnel exposure to ambient concentrations of airborne contaminants remain within health based exposure indices and that personnel are alerted to hazardous releases.
- Property Protection – This protection is provided to detect and avert situations that could lead to fires or explosions. In some cases the intent may be to control other means of property loss, such as excessive corrosion due to airborne contaminants.
- Regulatory Monitoring – To ensure or prove that the concentrations of airborne contaminants are being kept below regulated levels for personnel safety or community impacts.
- Community Impacts – It may be useful to put detection equipment in place to detect or avert large releases that could pose a hazard to a large portion of the facility, environment and/or the public outside the fence line.

There are two motives for use of gas and vapor detection that should be avoided. Firstly, gas and vapor detection should not be used in lieu of good facility design and preventive maintenance. When a new use for gas detection has been proposed, first ask, "Have we done everything possible to minimize the potential for a release to the atmosphere?"

The probability of detection success in the event of a release may be very low due to the number of factors that the gas detection designers can neither account for nor control. The exact opposite is true for the designers and maintainers of the containment (operating) systems, as most of the conditions that the containment system may be exposed to are identifiable and predictable. As a result, there is a great opportunity to reduce the probability of release through proper equipment selection/design, preventative maintenance, detection and aversion of impending failure modes, or by controlling failures before they can get to atmosphere (such as the use of tandem or double seals on a pump). In many cases, the degree of protection provided per dollar spent on preventing recognized release events will be much greater than that which can be provided by gas detection.

Secondly, avoid using gas detection systems to satisfy individuals or groups of personnel. This is not only limited to concerned employees, but also includes managers who feel the need to show everyone that they have done "something" to make improvements following a bad experience or incident.

Again, one should first pose the question, "Have we done everything possible to minimize the potential for a release to atmosphere?" Secondly, the management program should be reviewed to see if personnel requests fit in to it or if the program needs updating. Finally, the decision can be made as to whether the installation should be pursued based on the management program. To do otherwise, leaves the company susceptible to inadvertently:

- Addressing a recognized hazard with a fix that may have a very low probability of success compared to addressing it directly (i.e., prevention)
- Fostering a false sense of security amongst personnel, as a result of:
  - Employing a quick fix that may not provide the level of protection afforded by a well engineered system.
  - Creating an unusually high level of protection in one area, which may lead employees to believe that this level of protection has also been provided elsewhere.
- Prioritizing available funding from more critical safety needs

2.3 WHAT DO WE WANT TO DETECT?

The answer to this question is not as simple as, "We want to detect all leaks", as gas detection systems are like all many other manmade devices; they only work properly if the application is within their capabilities to perform.

In order to assure that gas detection systems are effectively utilized, the company needs to identify the characteristics of the release(s) that it is concerned about in as much detail as practical. In many cases, these characteristics may involve ranges in magnitude (1 lb/minute vs. 10,000 lb/min), timing (normal vs. after hours) and environmental conditions (winter vs. summer).

Taking the time to proactively address this up front will allow the company or facility to make its gas detection decisions based on objective technical merits. If used properly, this information will help to:

- Verify that the releases of concern can actually be detected with a relatively high probability of success. If this is not the case, it will help everyone feel more comfortable in moving towards preventive or other detection technologies to better address the situation.
- Assure that the planned response to the release of concern is appropriate. As an example, it may be totally acceptable and safe for properly equipped personnel to enter a toxic atmosphere to manually isolate a small release. However, it is probably not acceptable to send personnel into a 10-ton release of flammable gas in hopes of isolating it before it ignites.
- Understand the limits of the detection provided by the gas detection systems. This will minimize the potential for over reliance on the gas detection system and foster an atmosphere where personnel apply appropriate degrees of awareness and precaution.

2.4 WHAT ACTIONS DO WE EXPECT TO UNDERTAKE IN THE EVENT OF A RELEASE?

As described above, it is important that gas detection systems only be used where the actions that they will trigger are appropriate for the type of release that is likely to have been detected. In answering this question, three things need to be looked at:

- Why was the detection installed?
- What are we trying to detect?
- Are the containment systems and personnel equipped to resolve the situation safely and without incident?

If the system or personnel are not adequately equipped to resolve the situation the gas detection system under consideration should not be installed until an appropriate and safe means of dealing with the release has been identified and implemented. To do otherwise, may cause personnel in relatively safe locations to move towards a hazard that they are not equipped to safely handle.

In most cases, identifying the appropriate means of handling the release will come down to matching the means of isolation to the hazard of the release. Means of isolation may include:

- Manually applied plugs, patches and clamps
- Manual block valves or power operated valves with local controls
- Power operated valves with remotely located controls
- Power operated valves that are automatically activated by the gas detection system
- Emergency shutdown systems on rotating, reciprocating, fired and other equipment. These systems may be locally, remotely or automatically operated.
- Additional ventilation
- Deluge system

Another choice for dealing with the release is to divert the process or storage material to a safe location by utilizing a depressurizing system, rapid inventory transfer system, or even to an adjoining process loop or train. The means of accomplishing this (locally, remotely or automatically) would still have to be commensurate with the risk to personnel.

Another choice for addressing releases while keeping them safely within the capabilities of personnel may be the use of vapor mitigation or dispersion systems. Water based vapor mitigation systems, such as water spray systems/curtains and remotely operated fire water monitors can be effective on water soluble materials such as hydrofluoric acid and ammonia. However, their use on non-soluble materials is limited unless an appropriate additive is utilized. Vapor dispersion systems on the other hand do not rely on solubility of the containment, but utilize a medium (usually water) to improve the mixing of the contaminant with fresh air, thus driving its concentration down below hazardous levels. In many cases, the use of these vapor dispersion systems is limited to small or very well defined releases; they are usually operated remotely or automatically.

Where none of the above-mentioned means of dealing with the release is practical, two applicable choices remain. The first is to utilize the gas detection system alarm to evacuate the area and allow the situation to progress naturally. In some cases, this may actually be the best and safest choice from all perspectives (personnel safety, property damage, environmental impact and public exposure). If allowing the situation to progress naturally would not be acceptable, serious consideration should be given to the second applicable choice, redesign the containment system to curb the potential for releases or to reduce the size of potential releases to that which can be safely addressed.

A word of caution about the use of automatically activated release control systems is warranted. Gas detectors are subject to mechanical and electrical failures that could initiate a false trip of the system. They may also be subject to false trips due to improperly selected set points, interferences from exposure to non-target gases, or releases from areas that are not associated with the control system. As such, the use of automatic release control systems should be limited to those situations warranting this level of response and to those process and storage systems for which it has been determined that activation will not result in unsafe operating conditions.

2.5 HOW MUCH SHOULD WE SPEND ON DETECTION?

The amount of money spent should be optimized to produce the needed results with minimal expense.

The only way to accomplish this is by managing gas detection systems from cradle-to-grave by establishing the procedures or standards needed to show effective utilization. This program should answer all of the questions discussed above and should provide guidance on:

- When preventive measures are to be taken before gas detection may be considered.
- Why the various types of detection are to be used.
- The type/size of release sources to be protected.
- Ability to detect limits on release.
- Requirements for release control (before a system may be installed) including procedures (operating, maintenance and emergency); personal protective equipment; emergency isolation, diversion and shutdown systems; vapor mitigation and dispersion systems; and the use of inherently safer designs.
- General guidance on detector selection, detector placement, set points and alarm management

This program should be coordinated with and incorporated into the Process Safety and other risk management programs of the company. It is particularly important that the gas detection program be coordinated with the company's management of change process to assure that:

- The use of gas detection is expanded appropriately when new equipment, processes or materials are introduced.
- Existing gas detection systems are decommissioned when the previously protected equipment processes or materials are eliminated.

3
Determining Where Gas Detection May or May Not be Beneficial

The decision to design and install gas detection is based on the risks and culture present within the facility, as there are only a few cases where gas detection is legally required.

This chapter will provide guidance and concepts that may help when evaluating a situation and determining where gas detection may or may not be beneficial.

3.1 ASSESSING WHERE GAS DETECTION MAY BE BENEFICIAL

The determination of when and where gas detection is to be provided is predominantly based on the release scenarios and risks present at the facility. Depending upon the complexity and size of the facility, the assessment of these risks may be completed informally or through the use of more rigid risk evaluation tools such as:

- Process safety methodologies such as preliminary hazard analysis, HAZOPs and layers of protection analysis.
- Fire protection risk analysis methodologies, such as those published by the National Fire Protection Association (NFPA 550 and 551) and the Society of Fire Protection Engineers

Once the risks of the various release scenarios have been identified they can be compared to the facility's risk tolerances, possibly using its process safety management (PSM) risk matrix as a basis. In completing this evaluation, gas detection would be considered for those scenarios where the:

- Existing risk exceeds the facility's risk tolerances; and
- Further analysis indicates that the scenario's risk would fall to within acceptable levels if gas detection were present.

A positive response to the second requirement is important, as there would be little benefit to providing gas detection if it does not return a suitable risk reduction. In evaluating this risk reduction, it is essential that the facility be realistic in what it expects to accomplish once a release has been detected. As an example, if the facility estimates that it could bottle in a flammable release using remotely operated isolation valves within a few minutes of receiving the alarm there may be a significant risk reduction. However, if it is going to take an hour to cut rate and manually isolate the release there may be little risk-reduction benefit in providing gas detection as the event is still likely to have reached steady state conditions and escalated to its natural (without intervention) consequences.

In comparing the before and after risks of a release scenario it is also essential that the facility take into account the emergency procedures that will be implemented upon receipt of a gas detection alarm. The reason for this is that a manual response to investigate and isolate a release may actually increase the potential for personnel to be exposed to the release or its accompanying fire/explosion if appropriate procedures or isolation measures are not available.

The risk posed by a release scenario is the combination of its probability of occurrence and its potential consequences. In determining the probability of occurrence the type of operation, operating conditions (temperature, pressure, etc.) and the equipment involved are all considered. In determining the consequences the size of the leak, the proximity of the leak to other equipment and populations (onsite and off site) and the speed with which the leak is likely to be detected without gas detection are all considered.

The risk review should evaluate the probability of release occurrence for all possible release scenarios at the facility, including those caused by:

- Human errors (i.e., failure to close a drain valve, incorrect valve opened)
- Active equipment failures (i.e., pump seal or bearing failure)
- Passive equipment failure (pipe gasket failure or pipe rupture)
- Sabotage or terrorism

In addition the risk review should take into consideration that certain areas of refineries, chemical plants and manufacturing facilities have historically higher probabilities of releases than others. These areas typically encompass:

- Open sample points, drains, and vents, especially where they are left unplugged or uncapped during normal operations
- Equipment that is opened for maintenance, particularly if it's opened while the plant/unit remains on line (i.e., strainer and filter housings)
- Seals on rotating equipment, such as pumps and compressors
- Atmospheric seals on sulfur recovery units
- Flange gaskets
- Tanks, due to overfilling
- Small bore piping in vibrating services

- Lines in corrosive or erosive services
- Plate and frame heat exchangers

The risk review will also need to evaluate the consequence of each release scenario, which may range from inconsequential to disastrous. The severity of the consequences will depend upon the following:

- The material leaked and its physical properties under process conditions.
  - The vapor pressure of the material at process temperatures is of particular importance as this will drive the degree to which vaporization and whether or not the leak is even detectable using gas detection.
  - The temperature of the material relative to its boiling point, auto-ignition temperature and flash points as this will affect the rate of vaporization and the degree of fire hazard posed by the release.
  - The storage or processing pressure of the material, as this will affect whether the material is released as a slow stream with little vaporization or as a fine spray with increased vaporization.
- The mass release rate which is a function of the material's vapor pressure, the operating temperature and pressure, and the cross sectional area of the opening.
- The proximity of the leak to in-plant personnel, ignition sources and property lines.
- The ability of personnel to detect the leak by direct observation or with process instrumentation.
- The ability to reduce processing rates or isolate the source once a leak is discovered.
- Atmospheric conditions, especially wind direction and velocity

Where necessary, a better understanding of the risk posed by a release scenario may be acquired by reviewing industry data on release and equipment failure frequencies and through the use of consequence modeling to evaluate the consequences of postulated releases, fires, and explosions.

Once the risk of release is understood, the guidelines and concepts included the following sections of this chapter can be used to further evaluate where gas detection systems should or should not be deployed.

3.2 SITUATIONS WHERE OTHER TECHNOLOGIES MAY BE MORE BENEFICIAL

Toxic and combustible gas detection may be utilized in numerous situations, but may not always be the best choice for alerting personnel to the presence hazardous atmospheres. The primary reason for this is that gas detection systems can only be laid out and configured (see Chapter 5) to maximize the potential for detection success, as opposed to guaranteeing detection success in most situations.

There are simply too many variables affecting the travel and dispersal of airborne releases to make the layout and configuration of gas detection systems an exact science. Therefore, one should always consider prevention and detection alternatives that may offer a higher probability of success in preventing the occurrence of unobserved releases.

It is generally more effective to address potential gas and vapor releases by reducing the probability for a loss of containment than by using gas detectors once gases or vapors have been released to the atmosphere. Adequate safeguards may be identified through the use of improved engineering standards and process hazard analysis methods, such as layers of protection analysis (LOPA) and fault tree analysis. Safeguards that may negate or reduce the need for gas detection include, but are not limited to:

- Seal-less designs for rotating equipment.
- Tandem or double seals on rotating equipment. These systems are especially effective where remote monitoring and alarming of the seal interstitial space has been provided.
- Adequately alarmed dry gas seals on compressors.
- Redundant seals, O-rings, etc. on equipment that has been designed with access panels, hatches, and covers for maintenance purposes.
- Safety instrumented systems, vibration monitoring systems or other applicable instrumentation to detect conditions that may lead to a failure of containment.
- Double walled tubing, hoses, and piping where practical.
- Increased piping wall thicknesses and corrosion allowances.
- Safeguards such as excess flow valves, automatic shut-down valves, and flow restricting orifices that may be used to reduce the duration and volume of a release.

Another thing to consider is that fixed gas detection is often installed on the assumption that if one is concerned about a gas being present in the atmosphere, one must use gas detection. However, this may not always be the case when the released material is in the liquid phase or includes a significant liquid component. In these cases, it may be more effective to detect a release based on the presence of its liquid component, because it is much easier to predict where this liquid will collect and travel (this will always be driven by gravity) than it is to predict where its vapors may travel. Liquid detection technologies that could be used in lieu of or in addition to fixed gas detectors include, but are not limited to:

- High temperature detection – These systems utilize temperature sensors placed in containment areas, trenches, and sewer systems where hot process fluids may run to or accumulate upon release.
- Low temperature detection – These systems utilize temperature sensors placed in containment areas, trenches and sewer system where cryogenic materials such as LNG may run to or accumulate upon release.

- Oil on water detection – These systems measure electromagnetic energy absorption to identify when the layer of hydrocarbon product on the surface of accumulated water has exceeded threshold thicknesses. These devices could be placed in containment areas, trenches and sewer systems.
- Liquid/liquid concentration – These systems measure microwave absorption of a liquid to identify the percentage of hydrocarbon within a water stream or vice versa. These devices could be placed in containment areas, trenches, piping traps, and sewer piping.
- Liquid level detection – There are a multitude of technologies available for monitoring liquid levels. These devices could be placed anywhere released materials may accumulate.

The use of liquid detection may also be more effective than gas detection in those cases where the released material has a low vapor pressure and its vapors are heavier-than-air. It may prove difficult to use gas detectors in these cases because the vapors may not reach the sensors (at alarm threshold concentrations) given the elevation at which they must be mounted for maintenance and reliability purposes.

Another non-traditional detection technology that could be considered is sonic leak detection. This fairly new technology utilizes microphones to listen for the audible signature of a gas release, and is not dependent upon the released gas contacting a sensor to initiate an alarm. However, this technology is only applicable to situations where gases are being handled or stored at relatively high pressures. This technology may also be limited by background noise levels (requiring more detectors).

3.3 SITUATIONS WHERE GAS DETECTION IS RECOMMENDED BY CONSENSUS OR MANDATED BY LAW

There are recognized codes and standards that may be helpful in determining if a facility either demands or may benefit from a gas detection system. In applying these standards and codes, it is essential that one understand when these documents are simply "best practices" that may be used on a voluntary basis and when they carry the weight of law and must be complied with.

In general, a code or standard cannot pose a legal requirement unless one of the following two conditions applies:

1. It has been specifically adopted into law by a federal, state, county or local government.

   As an example, the Commonwealth of Pennsylvania has adopted the International Fire Code (IFC) as its base fire code. Therefore, the requirements of this document must be complied with if the facility is located within Pennsylvania. On the other hand, NFPA 1, "The Uniform Fire Code," has not been adopted, making it a "best practice" from which the user can pick and choose guidance, except where condition 2 applies.

It is important to note that many jurisdictions will take exceptions and make additions to their adopted codes as part of their enabling legislation. Therefore, one may need to contact the code authority or obtain a copy of the enabling legislation to verify the jurisdiction's true requirements.

2. The document has been referenced by a legally adopted code or standard. In this case, the referenced document is only enforceable to the extent of the reference.

   As an example, the IFC contains no references to NFPA 318, "Standard for the Protection of Semiconductor Fabrication Facilities." Therefore, it is only a "best practice" in Pennsylvania and the user may pick and choose where they will apply its guidance. In another example, the IFC specifies that the fabrication of flammable liquid piping systems is to be in compliance with NFPA 30, "the Flammable and Combustible Liquids Code," making the piping requirements of this document legally binding in Pennsylvania. However, no other requirements of NFPA 30 would legally have to be complied with, unless they too are specifically referenced by the IFC.

The first step in determining a facility's gas detection requirements therefore, should be a complete review of applicable federal, state and local regulations and codes. This review will help the facility to identify the gas detection systems that must be provided in order to maintain legal compliance.

Once the facility has identified what systems are required it can then evaluate whether or not further protection is warranted. In doing so, the facility should consider consensus and insurance industry standards applicable to its operations and the basic guidance provided in Sections 3.1, 3.3, and 3.4. Table 3.1 provides examples of documents that may be helpful in evaluating the gas detection requirements of a facility.

Table 3.1 – Relevant fire codes.

Organization	Title	Summary
American Petroleum Institute	API RP 751, *Recommended Practice for Safe Operation of Hydrofluoric Acid Alkylation Units*, 1999	Each HF alkylation unit should have an effective leak-detection system. Such a system may include closed-circuit television, point sensors, open-path sensors, and other systems deemed appropriate for the unit.
American Petroleum Institute	API Std. 2510, *Design and Construction of LPG Installations*, 2001	A safety analysis should be used to evaluate the need for hydrocarbon detection.

Table 3.1 – Relevant fire codes (continued).

Organization	Title	Summary
American Petroleum Institute	API Publ. 2510A, *Fire-Protection Considerations for the Design and Operation of Liquefied Petroleum Gas (LPG) Storage Facilities*, 1996	Hydrocarbon vapor detectors may be provided for area surveillance, and are best used as a pre-fire vapor-detection early warning system.
FM Global (Property Insurance Organization)	FM 7-13, *Mechanical Refrigeration*, 1991	This type of detector would be used to alarm and actuate ventilation systems in existing ammonia refrigeration areas.
GE Global Asset Protection Services (Property Insurance Organization)	GAP.17.12.1, *Fire Protection For Electric Generating Plants And High Voltage Direct Current Converter Stations*, 1996	Carbon monoxide gas detection systems should be provided.
National Fire Protection Association	NFPA 318 *Standard for the Protection of Semi-conductor Fabrication Facilities*, 2006 Edition	Gas detection is required for toxic and highly toxic compressed gases. Detection levels are set at OSHA PELs for ambient or breathing zone releases and the NIOSH IDLH for exhausted enclosures.
National Fire Protection Association	NFPA 55 *Standard for the Storage, Use, and Handling of Compressed Gases and Cryogenic Fluids in Portable and Stationary Containers, Cylinders, and Tank*, 2005	Gas detection system for toxic and highly toxic compressed gases; monitoring the exhaust system at the point of discharge from the gas cabinet, exhausted enclosure, or gas room. Detection is set at: 1. The fail-safe valve shall close when gas is detected at the permissible exposure limit, short-term exposure limits (STEL), or ceiling limit by the gas detection system. 2. For gases used at unattended operations for the protection of public health, such as chlorine at water or wastewater treatment sites, the automatic valve shall close if the concentration of gas detected by a gas detection system reaches one-half of the IDLH (immediately dangerous to life or health).

Table 3.1 – Relevant fire codes (continued).

Organization	Title	Summary
		3. The gas detection shall also alert persons on-site and a responsible person off-site when the gas concentration in the storage/use area reaches the OSHA, PEL, OSHA ceiling limit, or STEL for the gas employed.
National Fire Protection Association	NFPA 5000™ *Building Construction and Safety Code*, 2006	A continuous gas detection system shall be provided for hazardous production material gases when the physiological warning properties of the gas are at a higher level than the accepted PEL for the gas and for flammable gases.
National Fire Protection Association	NFPA 560 *Standard for the Storage, Handling, and Use of Ethylene Oxide for Sterilization and Fumigation*, 2004	Indoor dispensing areas shall be equipped with a continuous gas detection system that provides an alarm when ethylene oxide levels exceed 25 percent of the lower limit of flammability (7500 ppm).
International Code Institute	*International Fire Code,* Ch 37, Sect 3704, Ch 27, Ch 18, 30, 35, 41	Outlines Gas detection requirements for toxic and highly toxic compressed gases and defines the detection level at the OSHA PEL.
National Fire Protection Association	*Flammable and Combustible Liquids Code*, 2003	Vapor detection systems shall sound an alarm when the system detects vapors that reach or exceed 25 percent of the lower flammable limit of the liquid stored. Vapor detectors shall be located no higher than 300 mm (12 in.) above the lowest point in the vault.
Compressed Gas Association Inc.	CGA P-32 – 2000 *Safety Storage and Handling of Silane and Silane Mixtures*	Requires the use of UV/IR detection to detect a release; this is due to the pyrophoric nature of silane.

3.4 SITUATIONS WHERE TOXIC GAS DETECTION MAY BE BENEFICIAL

In addition to considering the factors discussed in Sections 3.1 and 3.2, the toxic and other properties of the materials also need to be considered in evaluating the need for toxic gas detectors.

One important characteristic to be considered is the relative toxicity of the materials being handled at the site. In doing so, it is common for organizations to use varying detection requirements depending upon the degree of toxicity posed by the gas or vapor, where:

> **Toxic** is defined as a chemical having a median lethal concentration (LC_{50}) in air of more than 200 parts per million but not more than 2,000 parts per million by volume of gas or vapor, or more than two milligrams per liter but not more than 20 milligrams per liter of mist, fume, or dust, when administered by continuous inhalation for one hour (or less if death occurs within one hour) to albino rats weighing between 200 and 300 grams each.[1]

> **Highly toxic** is defined as a chemical having a median lethal concentration (LC_{50}) in air of 200 parts per million by volume or less of gas or vapor, or 2 milligrams per liter or less of mist, fume, or dust, when administered by continuous inhalation for one hour (or less if death occurs within one hour) to albino rats weighing between 200 and 300 grams each.[1]

Another important characteristic to be considered is the warning properties inherent with the materials being handled. These warning properties may include the gas' color, odor, taste, and — in some cases — its acidity and corrosiveness. Agents with poor physiological warning properties are characterized as those materials that are undetectable by human smell, taste, and sight and those materials which act to overwhelm or deaden the body's sensory mechanisms so quickly that they cannot be relied upon for detection.

In general, when the concentration required to trigger a physiological warning is higher than the accepted Permissible Exposure Limit (PEL) or other Occupational Exposure Limit (OEL), a gas detection system should be considered in the work area (*Odor Thresholds for Chemicals with Established OELs*[2] is one reference for these properties). Examples of materials having poor warning characteristics include:

> **Carbon monoxide (CO)** – Carbon monoxide is a colorless, tasteless, and odorless gas. Therefore, this gas has no physical properties to signify when it is present in an atmosphere, thus increasing the risk of encountering dangerous concentrations without warning.

> **Hydrogen sulfide (H_2S)** – Hydrogen sulfide is commonly described as smelling like rotten eggs at low concentrations, but deadens the sense of

smell at slightly higher concentrations. As a result, those exposed to this material may fail to retreat from dangerous concentrations of the gas because their senses were immediately deadened by exposure to higher concentrations or because they mistakenly believe that the gas concentration has subsided and is no longer a threat. H_2S in air may also reach explosive concentrations.

Inert gases such as nitrogen and argon – These gases can displace the oxygen molecules needed for respiration with no physiological warnings until it is too late. As a result, personnel could be exposed to such an atmosphere with no warning.

In addition to considering the characteristics of the toxic materials, the facility should also consider how it intends to manage and physically respond to toxic releases. By doing so, the facility will be able to develop coordinated policies and procedures concerning warning systems, evacuation routes, responder access routes, and mechanisms of control. It may also be helpful to consider the industrial hygiene monitoring requirements that an accidental release may trigger, as it may be impractical or unsafe to utilize portable monitoring equipment during such an event.

3.5 SITUATIONS WHERE COMBUSTIBLE DETECTION MAY BE BENEFICIAL

The primary consideration in evaluating the need for combustible gas detection is to determine if the area in question encompasses any sources of flammable vapors or gases. With this said, there are several types of operations that can usually be eliminated from consideration:

- Areas handling combustible liquids below their flash points, as there will be no vapors to detect.
- Areas handling combustible liquids, flammable liquids, or flammable gases at or above their auto-ignition temperatures, because these materials will ignite spontaneously upon being exposed to air.
- Areas handling lighter than air gases where no roof or deck is present above the release sources, as the footprint of any release is likely to be too small to practically detect.
- Areas where standing ignition sources such as fired heaters, boilers, and ground flares are in close proximity to the potential release sources.

The last of the areas discussed above always raises controversy as the natural inclination of many is to provide gas detection near ignition sources in the hopes of preventing the ignition of releases. However, this inclination should be resisted, as it is likely to have little, if any, benefit and could actually increase the risk to personnel because:

- Manually responding to a release in an area with standing ignition sources may place personnel in great peril. In this case, it will usually be safer for the facility to accept the ensuing fire while its personnel remain outside of the area performing their normal duties at the time of ignition.
- Interlocking the ignition source's controller with the gas detection system is usually impractical because of the time needed for the detection system, control system, and mechanical controls to react once a gas release has contacted the sensor. Even under relatively calm conditions (3 MPH winds) a gas release may be moving at speeds of more than 2 ft/second, leaving little time for intervention.
- Interlocking the ignition source's controller with the gas detection system may make the facility susceptible to spurious shutdowns due to false alarms. The use of a voted detection system (requiring more than one detector to trigger a shutdown) to avoid this problem will only increase the time delay associated with the interlocks making the installation even less practical.
- Even if the interlocks were successful in shutting down the fired equipment the convective currents will remain and are likely to pull the vapors into the equipment where they may ignite as a result of coming into contact with hot surfaces or being heated above their auto-ignition temperature.

It should also be noted that low vapor pressure flammable liquids may be difficult or impossible to detect with combustible gas detectors. This is particularly true in those cases where slow leaks or leaks with little atomization or spray are expected, because these vapors are quite heavy and have very little energy to drive their dispersion into the air. As a result, they tend to settle to the ground very quickly and may not reach detectable or alarm threshold concentrations at the heights (12–18 inches typically) at which the detectors must be mounted at for maintenance and reliability purposes.

At the other end of the spectrum, release scenarios involving high-vapor pressure materials and liquefied gases are especially conducive to detection with combustible gas detectors due to the large volume of vapor/gas generated by these materials. This propensity for generating large volumes of vapor also increases the risk of these scenarios making them more likely to be candidates for combustible gas detection during the risk assessment process. Examples of materials falling into this category include:

- Liquefied natural gas (LNG)
- Liquefied petroleum gases (LPGs)
- Liquefied gases or flammable/combustible liquids having a vapor pressure of 70 psi or greater at operating temperatures.

In general, it is impractical to provide combustible gas detectors to address every flammable release scenario in most refineries, chemical plants and manufacturing facilities. Therefore, their use is typically limited to relatively high risk applications were their presence may be of direct benefit. Depending upon the risk involved, the use of combustible gas detection may be beneficial where it is necessary to evacuate personnel, take emergency actions, or activate safety interlocks because there is the potential for flammable vapors to:

- Affect a normally or frequently occupied area or building.
- Be released from a source within a building in sufficient quantities to cause a significant explosion or fire.
- Engulf or enter a building housing ignition sources such as ordinary hazard electrical equipment and commercial/residential type heating equipment, boilers, etc.
- Be released and ignited in an unusually fire sensitive area such as those encompassing equipment constructed of low melting point or combustible materials, complex rotating equipment (compressors), and cable marshalling points.
- Accumulate or migrate into highly congested, obstructed, or semi-enclosed outdoor locations from which powerful explosions may emanate upon ignition. These areas are of particular concern because the explosions may be of sufficient magnitude to move and break process equipment, causing the rapid escalation of the incident.
- Accumulate or migrate into critical, unmanned, or remotely located facilities.
- Be released during large bore failures. Common examples of this situation include LPG truck, rail, and marine loading facilities that utilize hose lines.
- Entering an area beyond the owners control such as a public roadway or neighboring industrial facility.

3.6 EXAMPLE APPLICATIONS OF THE CONTINUOUS MONITORING SYSTEM

The application of continuous monitoring systems is perhaps best examined by industrial examples. This discussion begins with theoretical exposure scenarios, followed by specific applications.

3.6.1 Generalized Applications

In any manufacturing industry, a raw material has to go through four key stages before reaching its final product state. These stages are:

1. Raw material receiving/sorting/ storing
2. Raw material processing

3. Reaction/product generation
4. End form of product/packaging and shipment

Each of these four areas has inherent possible site exposures that are discussed below:

1. Raw material receiving/sorting/storing:

   In many industries the dock area for handling incoming shipments of raw materials can be a potential source for accident/exposure to working personnel. Many chemical industries use chemicals that are shipped in 55 gallon drums. The volatile nature of chemicals can lead to off-gassing of volatile vapor build up in the containers, which needs to be monitored. In storage areas, the presence of such drums in larger numbers can be a potential hazard. In some cases, the gases themselves may not be harmful to health, but in the presence of strong oxidizers they could catch fire and may present an explosion hazard. This also includes the semiconductor industry where cylinders containing high pressure toxic and highly toxic gases are received

   Continuous monitoring systems in such scenarios may be beneficial. Gas detectors to monitor for volatile components would ensure that the atmosphere does not exceed the LFL.

2. Raw material processing

   In manufacturing industries the incoming raw material is usually processed to some more usable form before being subjected to use. This process could be anything from changing the state of the raw material (i.e., using the gaseous form of a chemical species, which has been shipped in high-pressure containers in liquid form) to changing the physical structure of the raw material (i.e., pulverization of coal before mixing with other combustible materials). In semiconductor fabrications the greatest danger is when the cylinder valve is opened.

   Continuous monitoring systems in such cases may involve a gas detection system to ensure there are no leaks because of faults in process feed lines.

3. Reaction/product generation

   The processing area in a chemical manufacturing unit may utilize batch or continuous reactors. Various raw materials are fed in different physical forms, concentrations, and temperatures. A possibility of incompatibility between the interacting chemicals may occur.

   A common failure mode in such vessels includes valves connecting the feed line, the pressure relief safety systems, and the temperature monitoring sensors. For example, in high pressure boilers there is bleed off (blow off) valve which is opened while the boiler is being operated to blow off the sludge collected at the bottom thereby cleaning the boiler. Blocked valves and faulty pressure sensors may lead to explosion. Continuous monitoring of pressure and temperature in reaction vessels along with built-in safety measures and external shutdown mechanism will reduce the probability of such incidents. In the Semiconductor industries this also includes MOCVD reactors, Ion Implanters, Diffusion furnaces, and plasma etch processes.

4. End form of product/packaging and shipment

   Many products in chemical industries (e.g., sugar, pharmaceuticals) are further processed to either convert the state of the product to pellets, powder, etc. This involves drying operations that typically generate a lot of dust. The concentration of ambient dust, especially if it is combustible in nature, may need to be monitored to avoid explosion hazards.

3.6.2 Personal Direct Reading Monitors in the Refining Environment

The sulfur recovery unit (SRU) of an oil refinery handles large volume streams of concentrated hydrogen sulfide (H_2S). As a result, these units typically incorporate some degree of protection in the form of fixed H_2S detection.

In one case, a contractor working in an SRU was injured when he was overcome by H_2S. This incident was of particular concern because:

1. The unit had a relatively large number of grade-level point detectors that were laid out in more or less of a grid pattern. In fact, the location where this injury occurred was surrounded by detectors with the closest detector being no more than 20–25 feet away. All of these detectors were functional, but provided no indication of a release.
2. Several other contractors had passed through the same area and were in very close proximity to the contractor when he was overcome. Yet, no one had sensed or reported any unusual odors or indications of a leak.

This incident led the refinery to evaluate its H_2S detection practices and conclude that:

1. Even very small, localized H_2S leaks may be hazardous and that it was impractical to provide sufficient fixed detectors to meet all possible contingencies.
2. There were other units of the refinery handling H_2S streams with little or no fixed detection, compared to that provided in the SRU.

Due to this evaluation, the refinery has since implemented a policy requiring all personnel in process, storage, and transfer areas to wear portable H_2S

detectors. As a result, several potential incidents have been averted when personnel were alerted to the presence of H_2S by their personal detectors.

3.6.3 Perimeter Monitoring at an LPG Storage Facility

The following is an example of a comprehensive monitoring system deployed at a LPG cavern facility. This facility essentially consisted of a very large open field with many wellheads, pumps, and piping manifolds across the surface.

The open nature of the facility was a plus, providing very little confinement or obstruction, which greatly reduced the potential for explosions that could affect the public. However, there was still concern that flash fires could pose a threat to the public should a vapor cloud migrate onto the heavily traveled roadways and highways that encircled the facility.

As a result, an effort was undertaken to evaluate how gas detection could be utilized to alert operators in the event that a gas release threatened to enter a roadway. Based on this effort it was determined that:

1. Typical, small bore releases would dissipate and be well below their LFL before reaching the fence line.
2. Large bore releases could cross the fence line at concentrations above their LFL. However, this could be averted if the operators had sufficient warning to initiate the facility's emergency isolation systems and shut down its well pumps (i.e., reducing the system pressure).
3. Alarm set points needed to be well below the gas' lower flammable limit in order to provide operators with the time to respond.
4. Snow accumulations of 2 to 2-1/2 ft deep were common at the site, so detectors would need to be set approximately 3 ft. above grade to provide year-round coverage.

As a result of this effort, the facility was equipped with a combination of point and IR beam detectors. These detectors were arranged, as follows:

1. IR beam detectors were placed along most of the fence line and arranged to prevent them from crossing the facility's driveways in order to avoid frequent blocked beam alarms. These gaps in beam coverage were determined to be of no consequence, as computer dispersion modeling indicated that a vapor cloud from the target release could not pass through the gaps without overlapping one of the beam sets to either side of the driveways.
2. Beam detectors were not provided along the side of the facility bordered by a very steep, 20–30 ft tall incline. In this case, it was determined that the target releases were unlikely to make it up the incline at concentrations greater than the LFL and were more likely to travel along the base of the incline towards one of the other boundaries. As a result, point IR detectors were placed at each end of this incline (in addition to the end of an IR beam set).

3. The alarm set point of all the fence line detectors was set at the lowest practical setting, as false alarms from spurious, but normally expected releases could not occur given the setback distances.
4. Each well head was surrounded by point IR detectors laid out in a triangular pattern. The detector set back distances were selected to assure that the target release (smaller than that of the fence line detectors) would have time to settle or slump down, thus avoiding situations where the release could be carried over the detectors by their initial momentum.

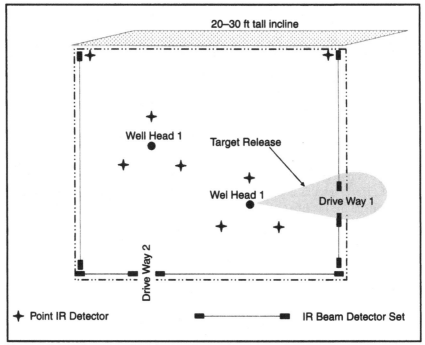

Figure 3.1 — Layout of Perimeter Detection at an LPG Cavern Facility

3.6.4 Exhausted Enclosures; Specific Example in the Semiconductor Industry, Specialty Gas Guideline[4]

The following (sections 3.6.5 – 3.6.7) is an example of a continuous monitoring system employed in the semiconductor industry. A simplified schematic of the system is presented in Figure 3.1, and the associated alarm level criteria for individual contaminants are listed in Table 3.2. Alarm set points and associated response protocols follow, along with a list of system features. Hazardous gas monitoring sampling points are described.

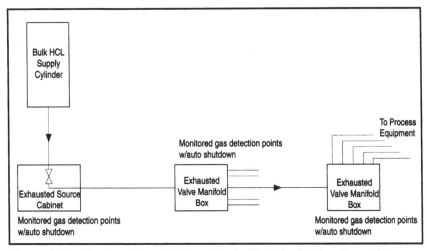

Figure 3.2 — Semiconductor Industry Toxic Gas Distribution Example

Table 3.2 – Example of Toxic Gas Control Requirement.

Name:	Formula	Gas Detection	Alarm Level	Auto shutdown
Ammonia	NH$_3$			
Argon	Ar			
Boron Trichloride	BCl$_3$	x	IDLH	x
Boron Trifluoride	BF$_3$	x	IDLH	x
Carbon dioxide	CO$_2$			
Carbon monoxide[1]	CO	x	IDLH	x
Chlorine	Cl$_2$	x	IDLH	x
Diborane	B$_2$H$_6$	x	IDLH	x
Dichlorosilane	SiH$_2$Cl$_2$	x	IDLH	x
Difluoromethane[1]	CH$_2$F$_2$	x	40% LFL	x
Fluorine[4]	F$_2$	x	IDLH	x
Germane	GeH$_4$	x	IDLH	x
Germanium tetrafluoride	GeF$_4$	x	IDLH	x
Hydrogen[1]	H$_2$	x	40% LFL	x
Hydrogen bromide	HBr	x	IDLH	x
Hydrogen chloride	HCl	x	IDLH	x
Hydrogen fluoride	HF	x	IDLH	x
Krypton	Kr			
Methane	CH$_4$	x	40% LFL	x
Methyl chloride	CH$_3$Cl	x	IDLH	x
Methyl fluoride	CH$_3$F	x	40% LFL	x
Neon	Ne			

Table 3.2 – **Example of Toxic Gas Control Requirement (continued).**

Name:	Formula	Gas Detection	Alarm Level	Auto shutdown
Nitrogen	N_2			
Nitrous oxide	N_2O			
Octrafluorocyclobutane	C_4F_8			
Octafluorocyclo-pentene	C_5F_8			
Octafluorotetrahydro-furan	C_4F_8O			
Oxygen	O_2			
Ozone *(greater than 12.5%)*	O_3	x	1xIDLH	
Perfluoropropane	C_3F_8			
Perfluorobutadiene	C_4F_6	x	IDLH	x
Phosphine	PH_3	x	IDLH	x
Silane	SiH_4	x^2	40% LFL	x
Silicon tetrachloride	$SiCl_4$	x	IDLH	x
Silicon tetrafluoride	SiF_4	x	IDLH	x
Sulfur dioxide	SO_2	x	IDLH	x
Sulfur hexafluoride	SF_6			
Tungsten hexafluoride	WF_6	x	IDLH	x
Xenon	Xe			

3.6.5 Example: Hazardous Gas Monitoring: Alarm Set points and Response Protocols

Gas Detection Alarm Set Points

Gas detection set points must be based on the hazard properties of the hazardous gas in question. For toxics, use IDLH; for flammable gasses, use LFL. If both are applicable, use the lower of the two.

The maximum low-level alarm shall be set based on half of the IDLH.

- Use half of the IDLH when available.
- For flammables, use 20% of the LFL.
- If no IDLH has been established, use the supplier's recommendation and available literature.

The maximum high-level alarm shall be set at the IDLH for toxics or 40% of LFL for flammables, depending on the detection range constraints.

Auto Shutdown

Auto-shutdown shall occur at a high-level alarm for all hazardous gases where gas detection is used. A formal risk assessment (example: PHA) will be required for any exception to be considered.

Detector Alarm

In the event of an alarm, visual/audible indication at a 24-hour security station is required. The ERT and detector owner are to be notified.

Low Level Detection

Visual/audible indication at a 24-hour security station is required. The ERT and gas systems engineering are to be notified.

High Level Detection

Visual/audible indication at a 24-hour security station and the process tool is required. Local evacuation of the area is also required. The ERT and gas systems engineering are to be notified. Auto-shutdown of the hazardous gas at the source should take place.

3.6.6 Hazardous Gas Monitoring: System Features

Life Safety Systems

Wiring interfaces between detection systems and annunciation/shutdown points will be via approved life-safety systems, utilizing fully supervised or fail safe circuitry.

Supervised Circuitry

Supervised systems and circuits must be capable of detecting and indicating the existence of an electrical open or short-circuit.

Response Time: Toxics

The response time for toxics should be within 30 seconds of initiation analysis.

Continuous Detection

Hazardous gas detection shall be continuous. The detection system is to be in continuous operation, and sampling is to be performed without interruption (see routine maintenance and sensor verification listed below).

Multiple Point Monitors

Cyclical analysis of multiple points may be used when monitoring within exhausted enclosures.

Cycle Frequency

When cyclical analysis is used, each point must be monitored every 30 minutes.

Routine Maintenance

Downtime for routine maintenance and repair should be less than one hour. In cases where downtime is required for more than one hour, a risk assessment must be conducted and an appropriate containment plan must be implemented.

Sensor Verification

All detectors must have an in-place sensor verification method, and routine sensor verification should take less than 30 minutes.

Interference

Interference by other chemicals should not result in alarm conditions.

Testing

Testing of the life safety hazardous system shall be conducted based on the manufacturer's recommendation. All alarming and indicating devices shall be tested to ensure proper response programming and component functionality. The method of testing shall be the same as initial acceptance testing.

3.6.7 Hazardous Gas Monitoring: Sample Points

Monitoring Locations

Sample points will be installed to monitor exhausted enclosures — the most likely places to indicate the presence of a leak should a gas line, fitting, or component fail.

Exhausted Enclosures

Exhausted enclosures include gas source cabinets, valve manifold boxes, tool gas boxes, and exhaust plenums.

Exhaust Dampers

Sample points will be located on the leak-source side of dampers installed in exhaust ducting.

Sample Point Locations

Sample points will be located a minimum of three duct diameters downstream from an enclosure (ideal design target is six to ten duct diameters). This will ensure adequate mixing of the target gas with exhaust air.

3.7 REFERENCES

1. "Occupational Safety and Health Standards,". *Code of Federal Regulations*, title 29, part 1910, subpart Z. Washington, D.C.: U.S. Government Printing Office, Office of the Federal Register, 2006.
2. **American Industrial Hygiene Association (AIHA):** *Odor Thresholds for Chemicals with Established Occupational Health Standards.* Fairfax, VA: AIHA, 1989.
3. **Intel Corporation:** *Intel Specialty Gas Guideline, REV 14.* Santa Clara, CA: Intel Corporation, 2002. [Unpublished Document]
4. **Union Carbide Corporation:** *Bhopal Information Center,* available at http://www.bhopal.com. Date accessed: November 8, 2006.

3.8 GLOSSARY

1. Immediately Dangerous to Life or Health (IDLH): Concentration defined by the National Institute of Occupational Safety and Health (NIOSH). IDLH values were based on the effects that might occur as a consequence of a 30-minutes exposure.
2. Lower Flammable Limit (LFL): The minimum concentration of a combustible substance that is capable of propagating a flame under specified conditions of the test.

4

Sensor Technology

4.1 INTRODUCTION

Many factors are involved when determining the type of gas and/or vapor detection equipment. Ultimately, one wants to choose the sensor that best provides protection of people and equipment. This chapter provides an overview of the different types of gas and vapor sensors available, the applicability and limitations of the equipment, and the factors to be considered when choosing gas/vapor detection sensors.

4.2 DESCRIPTION OF GASES AND VAPORS

4.2.1 Gas

A gas is a formless fluid, at 25oC and 760 mmHg, which expands completely to fill its container. Compared to solids and liquids, gases have a relatively low density and viscosity and are able to diffuse readily. Gases may be heavier or lighter than air. Understanding the characteristics and behaviors of gases targeted for sensing is essential for determining sensor equipment selection and placement. Examples of gases include hydrogen sulfide, ammonia and hydrogen. The particle size range of gases is typically less than 0.0005 um.

4.2.2 Vapor

A vapor is the gaseous phase of a substance that is typically a solid or liquid at 25°C and 760 mmHg. The process of evaporation changes a liquid to a vapor state and allows mixing with the surrounding atmosphere. The volatility of a substance will indicate how readily vapors may be emitted. Solvents such as benzene or methylene chloride will emit vapors readily. Volatile solids include naphthalene and mercury. The particle size range of vapor is typically less than 0.005 um.

4.3 AVAILABLE SENSORS AND HOW THEY WORK

4.3.1 Electrochemical Sensors

Gases that can be oxidized or reduced electrochemically can be detected by a fuel cell based electrochemical sensor. The target gas goes through either oxidation or reduction and causes a change in electrical potential in the cell, by chemically reacting with the electrode. The current produced is proportional to the concentration of the contaminant. While design varies depending on use and manufacturer, an electrochemical sensor generally operates using four key parts:

1. Electrodes: There are typically two to three electrodes (sensing, reference, and counter) made of a noble metal such as platinum. Electrodes in a single sensor are not necessarily made of the same material.
2. Membrane: The membrane allows the target gas to pass through to the sensing electrode while keeping water and particulates filtered out and preventing leaking or drying loss of the electrolyte. The membrane material and porosity will depend on the sensor type and manufacturer.
3. Electrolyte: This is typically an aqueous acid or salt solution that carries the ionic charge across the electrodes. The presence of this material can make the sensor sensitive to temperature and humidity (loss or gain of electrolyte). Manufacturers are able to design their electrochemical sensor to accommodate for this sensitivity.[1]
4. Filter: This allows the target gas to permeate to the sensor and filters out unwanted gases. The filter can be made of activated charcoal or other material to ensure that it is appropriately selective for the target gas.

Electrochemical sensor life varies depending on the design and the target gas for which it is intended. Some sensors, like those for oxygen or ammonia, are consumed in operation and will not last as long as others. The typical life expectancy of these sensors is around two years.

4.3.2 Infrared (IR) Sensors

Many gases have characteristic absorption lines (absorption signature) in the infrared region. An IR gas sensor works by measuring how much IR light passing through a gas is absorbed by the gas molecules. The concentration of the gas is proportional to the absorption. This type of sensor will detect gases with molecules that contain two unequal atoms. Noble gases and gases such as nitrogen, oxygen, hydrogen, and chlorine will not be detected.

Infrared sensors are available in traditional point detector designs and beam designs, which can provide detection over a wide area. Most IR sensors are a single unit that has a pump drawing samples from different points into it. It sequences through all the points in a few minutes.

4.3.3 Catalytic Bead Sensors

This sensor monitors combustible hydrocarbon compounds. A coil wire is coated in glass or ceramic which is coated with a catalyst on which the gas is oxidized. The coil is electrically heated to a temperature that allows it to burn the target combustible hydrocarbon compound. The heat liberated by the compound is proportional to the concentration of the gas present. This causes the coil to heat up and subsequently increases its resistance. The resistance is then measured, providing a signal to the device.

The performance of these sensors can be inhibited or impeded by:

1. Halogen compounds: the sensor may briefly stop working after exposure to these chemicals. The sensor will regain functionality after a short period of time (24–48 hours).[2]
2. Silicon-containing products: These materials can coat the ceramic bead and render it non-functional. Cleaning products and oil or lubricants that contain silicon can cause this problem.

Depending on the application of the catalytic bead sensor, its typical life expectancy is about three years.

4.3.4 Photoionization Detector (PID)

Photoionization detectors are typically used to detect volatile organic compounds (VOC) using ultraviolet (UV) characteristics of the target gas. These use a high energy ultraviolet lamp to ionize the gas. The resulting charged ions emit an electric current that creates an output signal on the gas detector. Gases have discrete ionization potentials and lamps can be selected based on the ionization potential of the target gas.

Because these lamps detect a variety of VOCs within the operating range of the lamp, it is important to understand the target gas or potential mixtures of compounds that may be present in the environment. Manufacturers have designed correction factors and scrubber connections to aid in measurement of specific target gases. Additionally, it is important to know the ionization potential of the target gas to ensure the correct lamp is selected for detection. For example, a 9.5 eV lamp will ionize benzene (ionization potential: 9.25 eV), but will not ionize formaldehyde (ionization potential: 10.67 eV). This would need an 11.7 eV lamp.

The life expectancy of the lamp varies depending on the type used, but is relatively short compared to other sensor types. These must be cleaned periodically, and are often best for handheld detection instruments rather than continuous use monitoring.

4.3.5 Thermal Conductivity (TC) Gas Detectors

Thermal conductivity Gas Detector compares the thermal conductivity of the gas with a reference gas (typically air). A heated thermistor, or platinum filament, is exposed to the ambient gas while another, acting as a reference, is enclosed in a sealed compartment. Gases having a higher thermal conductivity than the reference gas cause heat loss from the exposed element, while those with a lower thermal conductivity cause heat gain. Such temperature changes cause changes in electrical resistance which is measured using a bridge circuit.

4.3.6 Colorimetry (i.e., Honeywell Zellwegger Chemcassette®)

A dry reagent works as a medium to collect and analyze air to detect gas. When the reagent is exposed to the target gas, a color change is produced in proportion to the concentration of the target gas present. Air is drawn through a paper tape or solution. The target gas will produce a color change, which is measured by a photometer and is proportional to the concentration of the gas present.

4.3.7 Radon Gas

Portable or fixed detection systems have sensors that use ionization chambers to detect the radon concentration in gas passing through the unit. An ionization chamber will measure the number of ions within a gas filled enclosure between two electrodes when a voltage is applied between the two. These electrodes may be in the form of parallel plates or of coaxial cylinders. One of the electrodes may be the wall of the vessel itself. When gas between the electrodes is ionized, an ionization current, which may be measured by an electrometer or galvanometer, is created.

These measure alpha particles and may have filters attached in order to prevent the passage of radon progeny so that only Radon-222 is measured. These work in wide temperature, pressure and humidity ranges.

4.3.8 Laser Gas Analyzer Open Path Detection

This technology utilizes lasers in the near infra-red spectrum that operate at a specific absorption wavelength for the gas to be detected. A target gas has a specific absorption wavelength meaning they absorb a portion of the light energy emitted by the laser. A gas molecule, when struck by the laser, begins to vibrate causing a change in the laser beam's superimposed frequency. A receiver will detect this difference after the laser beam is reflected back and a computer analyzes the difference in the superimposed signal and the returning beam in order to gauge the target gas. This technology can be used in both portable and fixed applications.[3]

4.4 FACTORS TO CONSIDER WHEN CHOOSING A SENSOR

4.4.1 Target Gas/Vapor

When installing continuous air monitoring equipment, one must determine the type(s) of sensor(s) needed for a specific application. Several factors should be considered when choosing a sensor. One important factor is the target gas/vapor. Each sensor type (e.g., electrochemical, infrared, catalytic bead and photoionization) is typically recommended for specific detection applications. For example, electrochemical sensors are often used for detection of toxic compounds such as hydrogen sulfide and carbon monoxide. Conversely, catalytic bead sensors are suited for detection of combustible vapors/gases in concentrations below lower flammable limits (LFLs). One must also consider the gases/vapors that are not detected by a sensor. For example, infrared sensors cannot detect hydrogen. Additionally, some compounds can be detected by more than one sensor type. Manufacturers of gas monitoring equipment are often in the best position to recommend a sensor for a specific application.

4.4.2 Interfering Gases/Vapors

Some gases/vapors are known to interfere with detection of the target gas/vapor. The interfering gas/vapor may also permanently damage the sensor. One common interference example includes silicon compounds degrading the performance of catalytic bead sensors. Silicone compound vapors may cause readings of combustible gases to be lower than actual gas concentrations. Additionally, catalytic bead sensors may be damaged if exposed to atmospheres containing lead compounds, halogenated hydrocarbons, reduced sulfur compounds and acid gases. Manufacturers of gas monitoring equipment typically publish cross-sensitivity data for specific sensors. Such data describe gases/vapors that may interfere with a sensor's performance. Sensor manufacturers also typically provide lists of gases/vapors that may potentially damage a specific sensor. Such data should be reviewed before using a gas/vapor detector in any environment.

4.4.3 Expected Gas/Vapor Concentration Range

Expected gas/vapor concentration must be considered when choosing a sensor, since the different sensor types have varying detection ranges. For example, a photoionization detector and a catalytic bead detector can both monitor flammable gas concentrations; however, their detection ranges vary significantly. Catalytic bead detectors are typically used to measure a flammable gas/vapor concentration as a percentage of the LFL (Lower Flammability Limit) (i.e., typical range = 1 to 100% of the LFL). Photoionization detectors measure flammable gas/vapor concentrations at levels below 1 part per million (ppm). Consider the example of xylene. Xylene has a LFL of 1.1% or 11,000 ppm; xylene has an OSHA permissible exposure limit (PEL) of 100 ppm. In the case of xylene, a catalytic bead sensor would be appropriate if one were concerned about a potentially explosive air-xylene atmosphere. The lowest xylene reading possible with a catalytic bead sensor is 1% of the LFL or 110 ppm. A catalytic bead sensor would not provide information regarding exposures below 110 ppm. Therefore, a photoionization detector would be more appropriate if one were concerned about occupational exposure to xylene at concentrations around the PEL. A PID, however, may not provide meaningful data for concentrations below 1 ppm.

When measuring LFL concentrations within an atmosphere, one must consider the expected maximum gas/vapor concentration and oxygen concentration. Catalytic bead sensors will give false readings in atmospheres with gas/vapor concentrations greater than the LFL concentration.

4.4.4 Fixed vs. Portable Detectors

Fixed or portable detectors may be used to monitor potentially hazardous conditions. Portable detectors are typically used for emergency response and confined space entry applications. It may neither be feasible nor cost-effective to place fixed detectors in spaces that are rarely entered (e.g., sewer manholes, boilers, and tanks).

Fixed detectors are typically used in areas that are often occupied and continuously have potentially hazardous atmospheres, both indoor and outdoor. If occupants of an area need to be warned of a hazardous atmosphere, fixed detectors with alarms may provide the quickest notification capability. In such examples, one or more detectors may be tied to an alarm that notifies occupants to take action or evacuate.

Fixed detection is particularly useful in enclosed spaces that contain a potential release source or draws air from an area that could be exposed to a release.

These systems can also be used in the outdoor environment. However, the degree of reliability or confidence is greatly reduced due to the number of uncontrolled variables in this environment such as wind, temperature, and topography. Fixed detection in this environment may be used to monitor potential sources with higher probabilities of failure or having the potential for large releases. They may, however, be better utilized for protecting specific receptors, population centers and/or approach/access paths in this environment. Fixed detectors are also typically used to monitor conditions along a perimeter of an area (e.g. fence line monitoring).

In some circumstances, it may be appropriate to use both fixed and portable detectors. For example, a facility may install fixed detectors within an area that has potentially dangerous gas concentrations. As a supplemental safety measure, employees who work in such an area may also wear personal gas detectors with alarm capability. In such cases, detection by either the fixed detector or one of the personal portable detectors could trigger an alarm and/or evacuation from a hazardous area.

4.4.5 Personal Detection Equipment

While personal detection equipment is not specifically covered in this document, it is important to be aware of the availability of many of the above sensors that come in the form of personal detectors. Circumstances may arise in which it is not feasible to provide adequate area monitoring. Portable sensors must be used to verify that the concentration in an area is safe. Some situations that may require the consideration of personal monitoring equipment as either primary or complementary gas/vapor detection include:

- Outdoor work environment with variable wind and weather conditions
- Emergency response activities
  - Fixed system damaged/not working
  - Response location not in vicinity of fixed monitors
- Performing special/non-routine tasks for which fixed monitors are not available
- Confined space entries
- Maintenance and construction activities

Many of these sensing technologies also come with data logging capabilities. Not only can this equipment provide real time indications/warnings of gases/vapors present in the worker's area, data can be retrieved to better understand full shift, short term and peak exposures for employees while performing certain tasks or in certain work areas.

4.4.6 Point or Open Path Detection

When utilizing fixed detectors, one may choose between point detectors or open path detectors. Point detectors measure gas/vapor concentrations within a relatively small area and are often used to detect gas/vapor leaks near a piece of equipment. If point detectors are used to cover a large area, multiple detectors may be needed.

Open path detectors measure gas/vapor concentrations within a relatively larger area and may be used in place of several point detectors for some applications. They consist of an infrared source and receiver. The light source emits an infrared beam along a straight-line path to a receiver. The receiver measures the infrared light. The open path detector monitors gas/vapor conditions within the area along the light path. If a gas/vapor cloud obstructs the infrared beam, the receiver will detect a reduced-intensity infrared beam. The receiver will then generate a signal indicating the presence of a potentially dangerous gas concentration along the light path. Open path detectors cannot be used to measure specific gas concentrations along a light path. For example, a small gas cloud with a relatively high gas concentration can produce the same measuring signal as a large gas cloud with a relatively low gas concentration. However, in this example, either scenario could be a potentially dangerous condition. Further investigation with another gas/vapor sensor would be needed to quantify the gas concentration along the light path.

Although open path detectors may be used to monitor conditions across a relative large area, it becomes more difficult to precisely locate the gas/vapor cloud as the distance between the light source and receiver increases. Weather conditions can also disrupt open path detectors, resulting in a loss of detection during periods of fog, snow, and rain. Great care must also be taken in beam placement to assure that expected vehicle parking/stopping, pedestrian travel, storage and/or vegetation growth will not interfere with the beams on a regular basis. For the best use of open path detectors, manufacturers should be consulted prior to installation.

4.5 SENSOR PERFORMANCE VARIABLES

All sensors have various capabilities and limitations. The following reviews different types of performance variables that should be considered when selecting the proper sensor for the desired application.

4.5.1 Response Speed

When choosing a sensor, it is important to understand the amount of time a sensor will take to collect, analyze and provide feedback for the desired concentration of the gas or vapor. Some detection devices actually have a built in time delay to minimize false alarms during events that may cause an unusual but temporary spike in the concentration of a gas or vapor. This feature may or may not be appropriate based on the flammability or toxicity of the substance being monitored.

4.5.2 Measurement Range/Operating Range

The sensor being chosen must be reviewed for its ability to detect a gas or vapor within the minimum ranges required for the application for which it will be used.

4.5.3 Sensitivity

When considering the use of gas/vapor detection sensing equipment for the protection of personnel and equipment, one must know what the minimum concentration of the contaminant is acceptable and ensure that the sensor can provide that level of detection.

4.5.4 Oxygen Requirements

Many sensors require oxygen in order to operate properly. When selecting a sensor, the need for oxygen must be taken account, along with its operating environment and the intended use.

For example, if oxygen concentrations are deficient or enriched, catalytic bead sensors will give false readings. Additionally, electrochemical sensors will not function properly without a small amount of oxygen.

4.5.5 Interference

Different environmental equipment-specific factors can impair the ability of a sensor to accurately respond to the gas/or vapor for which it is targeted. These factors include:

1. Gas/vapor and chemical mixtures: It is important to understand the complete mixture of gases, vapor and/or other chemicals (solid, liquid aerosol) that may be present in the environment being monitored and how these different materials will impact the performance of the sensors being selected. For example, electrochemical carbon monoxide sensors can have interference and false readings due to the presence of hydrogen sulfide, sulfur dioxide or hydrogen gas. Some sensors and their associated equipment may be sensitive to and damaged by corrosive materials.
2. Dust and dirt: The presence of dust and/or dirt on the optic portion of an infrared sensor system can cause interference and impair response. Some systems are designed to self-adjust to compensate for dust build up.
3. Temperature: Some sensors can have false readings if they are operated outside a specified temperature range. Hot and dry environments can shorten the life of an electrochemical sensor. Erroneous readings (either high or

low) may occur when electrochemical sensors are operated outside their specified range. This is an important consideration for indoor and outdoor installation. The user must explore the impact of temperature extremes both daily and seasonally. Understanding the behavior of target gas/vapors under certain temperature conditions is also critical.

4. Humidity: Climatic or operational conditions that create a humid environment can impact certain sensors such as electrochemical, infrared, photoionization, and thermoconductive. Depending on the sensor, false readings, decreased response or sensor damage may occur. The user must evaluate the environment, area operations, and housekeeping/maintenance activities for the potential of creating condensation/humid environments before selecting sensors that are vulnerable in humidity.

5. Air speed: Some sensors, such as catalytic bead, operate on air diffusion. Air speed that is too great can cause a disturbance to normal air flow in and out of housing and/or create turbulence causing differential cooling between the sensing components. In windy outdoor environments or extractive ducts, false alarms could occur if the sensor/housing is not properly placed or protected.

6. Flooding and snow cover: Detectors must be located so that expected flooding or pooling of water or accumulations of snow will not damage the sensors or preclude a release from contacting it.

7. Life span of the sensor: the life span of the sensor can be impacted by any of items listed above and will vary depending on the manufacturer. Some sensors that are exposed to constant, although low, concentration of certain contaminants will have a decreased life span or may incur damage. Understanding of the chemical mixtures in the environment, sensor vulnerabilities and diligent maintenance and inspection programs will help to ensure maximization of sensor lifespan and function.

8. Ease and frequency of calibration: Calibration needs will vary depending on the sensor and the manufacturer's recommendations. Decisions of what type of sensor to use may be partially based on the level of ease and frequency required of maintenance and calibration of the equipment. Some sensors, like infrared, may be relatively low maintenance while others may require more attention and functional knowledge.

9. Replacement cost: Understanding the life span, maintenance requirements and durability, and replacement of the equipment will help the user weigh out the costs and benefits of choosing a certain sensor. Evaluation of the appropriate sensor may include the cost of replacement vs. the cost of maintenance, repair and employee training for operation of, perhaps, more complex, yet equally effective sensors.

Table 4.1: Sensor Selection Matrix

Sensor/Detector	Advantages	Disadvantages	Typical Analyte(s)
Electrochemical	• 2-3 year sensor life. • A variety of systems are available. • Low limit of detection. • Do not get poisoned. • Generally high degree of selectivity.	• Use in hot and dry areas can result in shortened sensor life. • Gases like hydrogen, hydrogen sulfide, & sulfur dioxide can cause interference with certain sensors (filters can be applied). • Sensor life varies (1–3 years).	Carbon monoxide, hydrogen sulfide, chlorine, oxides of nitrogen, ozone, sulfur dioxide, oxygen, ammonia, some organics.
Infrared	• 3–5 year or higher source life. • Low limit of detection. • Direct read. • Analyzes many contaminants. • No contact with the gas. • No minimum level of oxygen required. • Relatively maintenance free. • Manufacturer can design to eliminate temperature fluctuation sensitivities.	• Overlapping absorption lines causes interference. • Can only monitor non-linear molecules. • Can be affected by humidity and water. • If optics are coated by dust or dirt, response may be impaired.	Hydrocarbons, aromatics, methane, ketones, ethers, and alcohols.
Catalytic Bead	• Small and rugged. • Simple calibration and maintenance. • Operates in a wide temperature range.	• Can be poisoned by trace amounts of substances containing silicon, sulfur compounds, chlorine and some heavy metals. • Halogen compounds and Freon can cause sensor inhibition (i.e., used in fire extinguishers). • High or continuous concentrations can burn out or permanently damage (crack) sensor which may not be detected until the next calibration. Following recommended calibration frequencies is extremely important.	Combustible hydrocarbon gases.

Table 4.1: Sensor Selection Matrix (continued).

Sensor/Detector	Advantages	Disadvantages	Typical Analyte(s)
Photoionization	• Fast response for many volatile organic compounds. • Can select different lamps for different compounds.	• Detector is non-specific and the user must know what is in the air and be able to characterize interference. • High humidity can decrease the response. • Lamp requires frequent cleaning. • Best for portable monitoring equipment.	Aromatic compounds, benzene, amines, ammonia, ethanol, acetone, acetylene, formaldehyde, methanol.
Thermal Conductivity	• Does not require oxygen to operate. • Can be used for a variety of gases.	Large amount of power is required to heat sensors. Sensors must be mounted in a flameproof enclosure. Cannot be relied on to measure gases with thermal conductivity to air ~ 1 (i.e., ammonia, carbon monoxide, nitric oxide, oxygen and nitrogen). Water vapor may cause interference.	Helium, hydrogen, methane, and neon gases (Thermal conductivity to air >1). Argon, butane, carbon dioxide, ethane, Freon/halon, hexane, pentane, propane, xenon, water vapor (Thermal conductivity to air <1). System not reliable for these contaminants.
Flame Ionization	Can be configured to respond to specific materials.	Requires hydrogen gas to operate.	Flammable compounds.
Colorimetry	• Contaminant specific. • Can test for a wide range of contaminants. • High sensitivity to non-reactive gases.	• Separate tape or reagent is required for each class of compound. • Can be expensive to install and maintain.	Sulfur dioxide, hydrogen sulfide, oxides of nitrogen, ammonia, chlorine, formaldehyde, isocyanates, phosgene, hydrazines, mineral acids, hydrides.
Laser Gas Detection	• Wavelength utilized is specific to target gas. • Quick response time (1s). • Long measurement path(~1 km). • Wide range of detection (0.1–1000 ppm for certain gases). • No degradation of sensing media.	• Only one gas measured per instrument. • Heavy dust, fog or steam can block the beam, however the reference and detection beam will be impacted equally. • Solid objects will block the path.	Hydrogen fluoride, methane, hydrogen sulfide, ammonia, hydrogen chloride, carbon dioxide.

In some cases, other leak detection technologies may be used in lieu of or to supplement combustible gas detection. Two typically utilized technologies include:

1. Low temperature detection – These systems utilize temperature sensors placed in containment areas, trenches and sewer system where cryogenic flammable materials such as LNG may run to or accumulate upon release. These sensors are then tied into a data processor that initiates an alarm when an unusually low temperature is detected.
2. Oil on Water Detection – These systems typically measure electromagnetic energy absorption to identify when a hydrocarbon product has entered a water only system. This technology may be utilized to alert operators when liquid phase releases are expected.
3. Liquid/liquid concentration – These systems typically measure microwave absorption of a liquid to identify the percentage of hydrocarbon within a water stream or vice versa. This technology may be utilized to alert operators when liquid phase releases are expected.
4. Sonic leak detection - This technology utilizes microphones to listen for the audible signature of a high pressure gas release. The advantage of these systems is that the released gas need not contact the sensor to initiate an alarm. However, this technology is limited to gas phase releases only, relatively high pressure systems and by background noise levels.

At the time of writing, no companies are known to use any of these technologies as a replacement for gas detection where it has been recommended. However, the oil in water technologies do provide a viable alternative in those cases where some type of detection is desired for environmental or safety purposes, liquid phase releases are expected and gas detection is not suitable.

4.6 REFERENCES:

1. **Honeywell Analytics:** *Gas Book.* Morristown, NJ: Honeywell Analytics, 2002.
2. **Chou, J.:** *Hazardous Gas Monitors: A Practical Guide to Selection, Operation, and Applications.* Columbus, OH: McGraw-Hill, 1999.
3. **Boreal Laser, Inc.:** *Gas Detection With a Laser Tutorial.* Spruce Grove, AB, Canada: Boreal Laser. Available at http://www.boreal-laser.com/docs/gasdetectionwithlaser.pdf. Accessed July 24, 2008.

4.7 GLOSSARY:

1. Lower Flammable Limit (LFL): The minimum concentration of a combustible substance that is capable of propagating a flame under specified conditions of the test.
2. OSHA Permissible Exposure Limit (PEL): OSHA sets enforceable permissible exposure limits (PELs) to protect workers against the health effects of exposure to hazardous substances. PELs are regulatory limits on the amount or concentration of a substance in the air. They may also contain a skin designation. OSHA PELs are based on an 8-hour time weighted average (TWA) exposure.

5

Approaches to Detector Placement and Configuration

The purpose of this chapter is to provide guidance on how to place detectors and configure detection systems once it has been determined that they are needed (see Chapter 3, "Determining Where Gas Detection May or May Not be Beneficial") and appropriate detectors have been selected (see Chapter 4, "Sensor Technology").

It is important to understand that the guidance provided in this chapter is intended to help the user maximize the potential for detection success in the event of a release, and that the detection of a release cannot be guaranteed.

5.1 GENERAL GUIDANCE FOR DETECTOR PLACEMENT AND CONFIGURATION

There are several methods for laying out gas detectors that should maximize the potential for detection success in the event of release. These include:

- Source monitoring
- Volumetric monitoring
- Enclosure monitoring
- Path of travel and target receptor monitoring
- Perimeter monitoring

Placing detectors in a random grid pattern is typically less effective than utilizing the placement methods described above because:

- Even under low wind conditions, releases may travel 100 or more feet in a matter of 15–30 seconds. As such, there is very little additional notice gained by placing detectors throughout the area over using the methods listed above.
- Detectors located in close proximity to release sources are less likely to see a release due to momentum affects which can propel it over, under or generally away from the sensors. The placement methods listed above avoid this by providing sufficient room between the release sources and

the detectors to allow the gases/vapors to settle downwards, increasing the probability of them contacting a sensor.
- The wiring cost of providing grid detection may be significantly higher than that of the methods described above, this is particularly true where:
  - The area has limited existing structures upon which the detectors, conduits, and cabling may be supported
  - IR or laser beam detectors can be utilized.

Wherever feasible, the placement of gas detectors should take advantage of topographical features and structures that may act to limit or control the dispersion of the target gas or vapor. This is especially true where heavier-than-air gases are of concern, as embankments, dike walls and valleys can be taken advantage of in order to maximize the potential of releases contacting a sensor and to minimize the number of detectors needed.

It should also be noted that it is seldom appropriate to use predominant winds as the basis for placing detectors in outdoor installations. These winds only occur maybe 50% of the time. However, it is appropriate and useful to take air currents into consideration when placing detectors for indoor installations. In doing so:

- Fine powders (commercially available for this purpose), air velocity meters, or fine strips of tissue paper can be used to identify how air moves through the enclosure to maximize the potential for a release contacting the detectors.
- Detectors should be considered for any appreciable dead air space that contains a potential release source.
- Detector placement within 12 inches of a corner (regardless of orientation) should be avoided, because the air velocities in these areas tend to be low or non-existent.
- Detectors can often be placed in or near the ventilation exhaust intakes, provided they are appropriately located with respect to the buoyancy of the target gas or vapor (within 12 inches of the roof or ceiling) and the air stream velocity limitations of the manufacturer are not exceeded.
- Detectors should be set off the walls and out in the more dominant air flows within the building, wherever practical.

In addition to air currents, it is important that the following also be considered in placing and configuring gas detectors:

- A "credible" release may not necessarily be detectable with any degree of certainty. As an example, a system designed to detect a full-bore seal failure on a compressor is unlikely to detect a small, yet potentially hazardous leak from a valve packing. Therefore, it is essential that personnel be made aware of these limitations through proper documentation and training.

- Process conditions and how they will affect the physical properties of the released material and its dispersion need to be considered. Released gases or vapors may behave differently than a review of the MSDS would indicate.
- Placing detectors close to a release source is seldom appropriate, as most process and storage system are under pressure. As a result, most releases will have a very small cross sectional area close to their source and will be carried over, under or away from nearby detectors by their momentum. Detectors should be set back from the release source to take advantage of the buoyancy characteristics of the vapor or gas in order to maximize the potential for contact with a sensor. Gas dispersion modeling may be helpful in determining appropriate set back distances.
- Interlocking gas detectors with automatic process controls and shutdowns may pose a higher risk in many applications than that posed by the potential gas release itself. This may make the process susceptible to upsets from inadvertent detector activations and/or induce severe process conditions that may threaten system integrity. Gas detectors should always be on a voted system (requiring at least two detectors to alarm before initiating any action) if they are to be interlocked with automatic process controls and shutdowns.

Regardless of the type and degree of gas detection provided, personnel must be taught that they cannot rely upon detectors to alert them to "all" releases. Gas detection must be thought of as a secondary layer of protection, with their personal actions being the first line of defense. These actions include:

- Adhering to safe work practices
- Using appropriate and recommended PPE
- Remaining cognizant of their surroundings at all times, paying particular attention to unusual or unexpected visual or audible cues that may be indicative of a release
- Responding to plant alarms and emergencies in accordance with established practices
- Adhering to the rule of thumb -"when in doubt, get out"

Personnel must also understand that the fixed gas detection should not be relied upon where area or task-specific monitoring is needed. Unless a sensor located in the specific area of concern, portable monitors should be used for task-specific monitoring ,such as confined space entry, hot work, and vehicle entry permitting.

5.2 GENERAL GUIDANCE FOR TOXIC GAS DETECTION

Toxic releases are problematic because their low concentrations of concern, as compared to flammable releases, can make even the most minor release a hazard.

The footprint of these minor releases can be very small, making it difficult to reliably detect them. This problem is further exacerbated by the fact that the sheer number of minor release sources present in most refinery, chemical, and manufacturing settings makes it practically impossible to comprehensively address all of them.

On the other hand, it is comparably easy to predict/control where personnel will normally enter an area, how they will traverse the area while performing rounds/assignments, and where they may spend significant amounts of time. As a result, the use of toxic gas detection is best focused on protecting frequently and normally occupied locations in order to meet the objective of alerting personnel to:

- Toxic and oxygen deficient atmospheres in buildings due to internal releases or the ingress of exterior releases. These systems may also be utilized to initiate the shutdown of the HVAC system and the process equipment that the buildings may house. See the guidelines for enclosure monitoring for further details.
- Releases that may affect commonly used access routes, normally occupied areas, emergency marshalling points, or the public. See the guidelines for travel path and receptor monitoring for further details.
- Large releases in the area of high-potential release sources. See the guidelines for source monitoring for further details.
- Releases that are affecting their immediate location. See the discussion bellow on the use of personal and portable detectors.

In the case of minor releases, it is generally more effective to address them by considering safeguards that will minimize the potential for a loss of containment or alert operators to conditions that are indicative of an impending failure. Adequate safeguards may be identified through the use of improved engineering standards and process hazard analysis methods, such as LOPA and fault tree analysis. Safeguards that may negate or reduce the need for toxic gas detection include, but are not limited to:

- Seal-less designs for rotating equipment.
- Tandem or double seals on rotating equipment. These systems are especially effective where remote monitoring and alarming of the seal interstitial space has been provided.
- Adequately alarmed dry gas seals on compressors.
- Redundant seals, O-rings, etc. on equipment that has been designed with access panels, hatches, and covers for maintenance purposes.
- Safety instrument systems, vibration monitoring systems, or other applicable instrumentation to detect conditions that may lead to a failure of containment.
- Double-walled tubing, hoses, and piping where practical.
- Increased piping wall thicknesses and corrosion allowances

- Alternative detection technologies, such as liquid level detection in drainage areas, temperatures sensors where cryogenic spills may occur, sonic leak detection, CCTV surveillance, etc.

Where the risk of a minor release from a specific source cannot be adequately addressed through preventive measures alone, the detector placement methods recommended above should be utilized. In addition, the following should be considered:

Use of personal and portable detectors - It is impossible to place fixed sensors to detect every minor release that may pose a hazard to personnel; this is especially true in outdoor applications. Personal and portable detectors, used as a supplement to fixed sensors, can provide a very effective means of warning personnel of leaks in their immediate area and outside the detection capabilities of their fixed equipment.

Personal detectors are now small enough that personnel can comfortably wear them on their person all day, and are cheaper and less likely to be set down or left someplace than traditional portable monitors. They provide a cost-effective means of providing toxic gas detection near the breathing zone of every person on the site.

Release likelihood - There is a tendency to place detection around release sources with high likelihoods of leak occurrence only. This practice may be inappropriate in the case of minor toxic releases, because the sheer number of release sources within a low likelihood area may actually make a release in this area more likely than in the high likelihood area.

As an example (using completely fictitious figures), the high risk portion of a unit has two potential release sources with a foreseeable failure rate of 1 X 10^{-3} leaks/year, resulting in it potentially experiencing one release every 500 years. The remainder, or low-risk portion of the unit, contains 100 potential release sources with a foreseeable failure rate of 0.5 X 10^{-3} leaks/year, resulting in the area potentially experiencing one release every twenty years. As a result, personnel are more likely to encounter a release in the "low hazard" area than in the "high hazard" area.

Height of the detectors and the number of levels - Detectors should be set at a height that will maximize the potential for contact with the vapor or gases as they disperse or to protect the breathing zone of individuals working in the area. In some cases it may be necessary to provide multiple vertical levels of detection where multiple materials or potential release types/orientation are present.

In the case of heavier-than-air gases, a minimum height of 18 inches (0.5 m) is recommended in outdoor installations to protect the equipment from dirt and splashing from rain or wash downs. In some places, where snow may accumulate to depths greater than 18 inches, it may be necessary to locate detectors higher than this. Also, where beam detection is utilized, any high points in grade between the sensors need to be considered, as wildlife, grass, and other vegetation may encroach upon the beam.

The use of combustible gas detectors, even where they are cross-sensitive to the target gas, is not recommended for toxic detection.

5.3 GENERAL GUIDANCE FOR FLAMMABLE DETECTION

The primary challenge in using fixed combustible gas detection is that it is impractical, if not impossible to detect minor releases, for the same reasons discussed under toxic detection. In addition, flammable releases will have much smaller detectable footprints than similarly sized toxic releases, because their concentrations of concern (essentially measured in parts per hundred) are much higher than those of toxic releases (measured in parts per million).

Moderate to large releases also pose a detection challenge because:

- The rate of area engulfment and vapor travel leave very little time for intervention by control systems or operators.
- The threat of flash fires and explosion make operations in the release area particularly dangerous for operating personnel and emergency responders.
- Explosions emanating from enclosed spaces or highly congested or obstructed portions of the plant can be quite powerful and result in widespread damage throughout the facility, thus spreading the incident.

Flash fires, on the other hand, typically do not lead to the immediate spread of the incident.

Due to these challenges, the objectives of most flammable detection programs in the process industries are limited to:

- Alerting personnel to the accumulation of combustible gases in buildings due to releases within the space or the ingress of exterior releases. See Section 5.5.1, "Applying the Volumetric Approach in Enclosed Buildings" for more details.
- Initiating the shut down of internal process streams/equipment and ventilation systems where combustible gases have accumulated in an enclosed space. Section 5.5.1, "Applying the Volumetric Approach in Enclosed Buildings"
- Alerting personnel to releases that may affect highly congested or obstructed areas of the plant from which powerful explosions may propagate. See Section 5.5, "Detector Placement for Volumetric Monitoring" for more details.

- Alerting personnel to large releases in the area of high-potential release sources. See Section 5.4, "Detector Placement for Source Monitoring" for more details.
- Alerting personnel to releases that may affect commonly used access routes, normally occupied areas, emergency marshalling points or the public. See Section 5.4, "Detector Placement for Source Monitoring," Section 5.8, "Detector Placement for Perimeter Monitoring," and Section 5.7, "Detector Placement for Path of Travel and Target Receptor Monitoring" for more details.
- Alerting personnel to releases that are affecting their immediate location. See Section 5.8, "Detector Placement for Perimeter Monitoring" and Section 5.7, "Detector Placement for Path of Travel and Target Receptor Monitoring" for more details.

Providing combustible gas detection in elevated locations, such as process unit decks and on offshore docks is not practical in most cases, because the number of factors at play precludes a reasonable degree of detection success. However, there are some situations where it may be practical to do so. These situations include:

- Areas sandwiched between two large solid decks or a large solid deck and roof.
- Docks handling products with high vaporization rates that may produce a large ball of vapor (propane, LNG, etc.).
- Decks that are substantially larger than the hazard area, so that the sensors may be placed on the deck and well away from its edge.

Where necessary, the source, volumetric or perimeter monitoring methods of detector placement could be utilized in these situations.

It is difficult, if not impossible to predict how vapors will behave as they approach the edge of a deck. In one case, the density of heavier-than-air gases could result in the vapors literally "pouring" off the deck, thus reducing the height (above the deck) at which a detectable concentration may occur. On the other hand, the turbulence at the downwind edge of the deck could result in the lifting or dilution of the vapors, making it difficult to estimate the alarm concentration to be targeted.

5.4 DETECTOR PLACEMENT FOR SOURCE MONITORING

Detector placement for source monitoring involves the placement of detectors around a potential release source with the intent of detecting leaks as they migrate away from it. This method of detector placement is applicable to the detection of moderate and large releases of flammable materials.

This method of detector placement may be used for toxic materials where the objective is to detect large or catastrophic releases that may migrate from the area. It should not be used where the objective is to detect small or moderate toxic releases that may affect those in close proximity to the equipment.

The first step in employing the source monitoring method is to determine if individual release sources should be considered individually or as one large release source. In general:

- If sources are well separated, each source should be surrounded by its own detection scheme.
- Where multiple point sources are within very close proximity to each other, perhaps within a 10 ft (3.3 m)–15 ft (4.5 m) diameter, they may be treated as a single point source emanating from the center of the circle.
- Multiple point sources that are spread out over an area may be handled by:
  - Identifying the limits of the area encompassing the sources
  - Identifying the credible and detectable release scenarios within the area
  - Placing the sensors around the perimeter of the area to achieve detection of the most difficult to detect release scenario, regardless of where it originates from within the area

The second step in employing the source monitoring method is to use computer modeling to evaluate the target release. This modeling should be used to evaluate practical orientations of the release to determine which will provide the least detectable vapor cloud (where feasible, horizontal releases with no impingement are generally the hardest to detect). Once the least detectable release is identified, the dispersion model should be evaluated to determine:

- The set back distance needed between the source and the sensors. The set back distance should be selected to assure that the cloud has settled down and that the sensors will come into contact with those portions having a concentration equal to or greater than the alarm set point.
- The allowable distance between sensors. In this case, the footprint of the cloud is evaluated to determine how wide the area of the cloud with a concentration at or above the set point concentration is at the sensor's height and set back distance. This value is then used as the maximum distance to be allowed between sensors. If desired, a safety factor can be provided for in establishing this distance by using a concentration that is slightly higher than that of the alarm set points during this evaluation.

Note: The vapor cloud dispersions depicted by computer models are merely a mathematical representation of averaged vapor clouds. The shape of actual vapor clouds will differ significantly from that depicted by these models. Therefore, the use of modeling in this step is limited to providing a rough idea of how a vapor or gas release will behave. Modeling under average weather and process conditions is considered adequate for this purpose.

There is no need to undertake in depth dispersion studies for the purpose of laying out gas detection.

The third step in using this method of detector placement is to lay out the sensors on a plot plan using the set back distances and sensor spacing limitations identified in the second step. In doing so, one should:

- Keep in mind that the sensors may be fixed point detectors, a portion of an IR or laser beam set, or a combination thereof. It is up to the designer to determine the type of sensors that will provide the most cost-effective means of achieving a reasonable probability of detection success.
- The dispersion of heavier-than-air gases will be affected by terrain and weather conditions. Therefore, these effects should be considered when laying out the detection.
  - Heavier-than-air vapors will tend to move towards low-lying areas. Consideration should be given to providing additional detection in these locations and/or focusing detection to take advantage of this behavior.
  - Depending upon the wind velocity, tall walls and steep inclines may cause some heavier-than-air gases to "fall out" as the air it is entrained in goes up and over the obstruction. This may result in reduced gas concentrations at higher elevations and "puddling" of the vapors (higher concentrations) at the base of the incline. In some cases, the designer may be able to credit the puddling or collection of vapors in an area to reduce the overall number of detectors or increase the spacing between them.
- IR or laser beams sets should be located so that frequently obstructed locations, such as roadways, are located within the gap between two adjacent beam sets. The allowable width of these gaps should be set so that the target gas plume cannot pass through the gap without overlapping one of the beam sets sufficiently to trigger an alarm.

As an example, the 10% LEL cloud shown in Figure 5.1 is expected to be 6 m wide at the height of the beams once it has reached them. In order to assure that the beam alarm setting of 10% LEL/m will be satisfied, this cloud must overlap one of the beam sets by at least 1 meter. As such the maximum spacing between beam sets should be no more than 5 m. If a wider gap were needed, a fixed point detector could be provided in the middle of the gap, allowing it to be enlarged to 10 m wide.

5.5 DETECTOR PLACEMENT FOR VOLUMETRIC MONITORING

This concept was initially developed by BP Oil for use on its offshore oil facilities, and is based on the concept that the greatest risk from a flammable release is that of direct damage from explosions, which have the potential to endanger

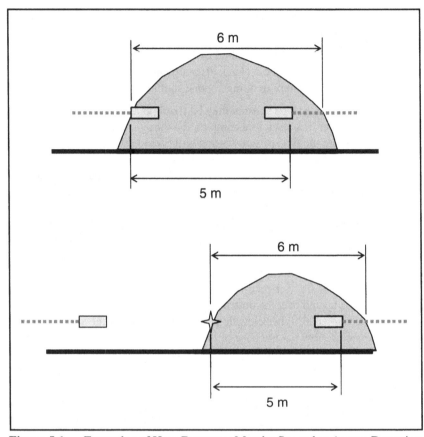

Figure 5.1 — Examples of How Detectors May be Spaced to Assure Detection Success

persons throughout the area and cause rapid incident escalation. This direct damage may be the result of piping and vessel movement due to pressure pulses or wind affects that can hurl debris or missiles throughout the area.

The volumetric monitoring approach utilizes a three-dimensional array of detectors (point, beam, or a combination thereof) to assure that a gas cloud, described in terms of a sphere of a specified diameter, cannot exist in the monitored space without contacting a sensor (See Figure 5.2, "Example of Volumetric Detection Concept" below). This method of detector placement does not consider small releases, which typically result in flash fires or inconsequential minor explosions only, based on the premise that it is impractical to reliably detect them with current technology.

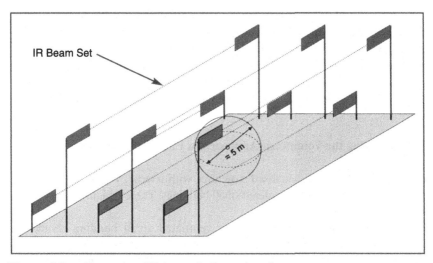

Figure 5.2 — Example of Volumetric Detection Concept

The diameter of the sphere used in laying out the detection array varies based on the type and size of space to be monitored, but is based on the premise that the following conditions must be present for an explosion to reach a significant magnitude:

- The flame front must reach a speed (through the vapour or gas) of approximately 330 ft/s (100 m/s) or greater.
- There must be sufficient space for the flame front to run up to its maximum speed. In the same way that an airplane needs a run up to reach take-off speed, the flame front of an explosion needs space in which to accelerate. In the case of alkanes, the most commonly handled type of flammable gas, a minimum acceleration distance of approx. 16–19 ft (5–6 m) is needed, given the geometry of most process facilities.
- A sufficient degree of blockage and/or congestion is needed, where,
  - Blockage would usually be defined as the ratio of the area blocked compared to the total cross sectional area in a given plane that is perpendicular to the direction of a flame front's travel.
  - Congestion would be indicative of multiple layers of obstruction.

Due to its focus on detecting conditions that can lead to significant explosions, the volumetric method focuses on the monitoring of highly obstructed and congested spaces where flammable gases may accumulate, as opposed to the source containment method, which focuses on areas from where large vapor releases may originate. In applying the volumetric approach to detector placement, it is irrelevant whether the releases originate from within or outside the space being monitored.

This approach is appropriate for use within buildings, partially enclosed structures, and congested or highly obstructed process structures that may be subject to internal or external flammable gas or vapor releases. It may also be applied where monitoring for large or catastrophic toxic releases should take place. However, the user would have to evaluate and determine the appropriate size of the detection target (defined by the diameter of a sphere shaped gas concentration) to be utilized.

5.5.1 Applying the Volumetric Approach in Enclosed Buildings

Buildings or other fully enclosed structures with internal volumes of less than 1,000 m^3 (35,000 ft^3) are considered small buildings. These buildings may be naturally or mechanically ventilated.

The detection within a naturally ventilated small building should be arranged to detect a spherical gas cloud of 4 m (13 ft) in diameter anywhere in the volume. If mechanical ventilation providing 6 ACH or greater is present it may be more appropriate to utilize the guidance for detector placement in enclosures covered elsewhere in this chapter.

In very small buildings, defined as those having less than 100 m^3 (3,500 ft^3), the minimum number of detectors to be provided should be based on the automatic actions to be initiated by the system, the detector voting philosophy of the facility and the layout of the space.

Buildings or other fully enclosed structures with internal volumes greater than 1,000 m^3 (35,000 ft^3) are considered large buildings. If an inerting system in compliance with NFPA 69 "Standard on Explosion Prevention Systems" is present, the detection should be capable of sensing a spherical gas cloud 5 m (16 ft) in diameter anywhere in the volume. If an inerting system has not been installed, the detection should be capable of sensing a spherical gas cloud 4 m (13 ft) in diameter anywhere in the volume.

If heavy gases are present, the three dimensional detection grid should include detection at approximately 12 inches above the floor.

Note: See Section 5.5.2 for semi-enclosed buildings and structures.

As indicated earlier, this methodology was initially developed for use in the offshore environment where the fully-enclosed volumes encountered are typically smaller and not as tall as those found onshore. As a result, its application in some onshore applications may require an unusually large and unnecessary number of detectors. The user may want to consider the following the guidelines for enclosure monitoring (Section 5.6, "Detector Placement for Enclosure Monitoring") or if the building is very large, employing the source monitoring method (Section 5.4, "Detector Placement for Source Monitoring"). In applying the source monitoring method in this case though, dispersion modeling should be utilized to consider releases oriented in all three planes in order to maximize the potential for detection success. In addition:

- Obstructed portions of the volume should be monitored to assure that a spherical gas cloud 5 m (16 ft) in diameter cannot exist without contacting a sensor.
- Detection should be provided in elevated locations if the potential for lighter-than-air or neutrally buoyant materials to collect there exists.

5.5.2 Applying the Volumetric Approach in Outdoor Locations and Semi-Enclosed Volumes

Open, outdoor locations do not require detection using this approach, as there is insufficient obstruction and congestion to generate a significant pressure pulse. Semi-enclosed volumes, such as beneath air-coolers, compressor buildings, etc., on the other hand must be evaluated on a case-by-case basis to determine if detection is warranted.

The first step in this evaluation is to determine the blockage ratio present in the semi-enclosed volume. The blockage ratio of a volume is determined by considering each volume to be a six-sided cube (in some cases, other geometric shapes may be more appropriate) then dividing the obstructed area present by the total area of the six sides. As an example:

- A large area of fin-fan air coolers is located atop a 15 m tall pipe rack. The area of the air coolers measures 11 m by 55 m.

  The total obstructed area of the cubic volume beneath the air coolers equals:

  Air coolers (11 m X 55 m) =   605 m²
  Ground Area (11 m X 55 m) =   605 m²
  *   1,210 m²*

  The total surface area of the 11 m X 55 m X 15 m cube equals 3, 190 m².

  Therefore,

  Blockage Ratio = 1,210 m² / 3,190 m² X 100 = 37.9 %

- The 11 m X 55 m elevated deck of a process structure sits 15 m above grade, consists of expanded steel and supports several exchangers. The obstruction presented by this deck includes:
  - The solid area of the expanded steel, which makes up approximately 30% of its square footage
  - Three exchangers having an aggreagate base area (when viewed from below) of 30 m²

The total obstructed area of the cubic volume beneath the air coolers equals:

Deck (0.30 X 11 m X 55 m) =	*182 m²*
Exchangers =	*30 m²*
Ground Area (11 m X 55 m) =	*605 m²*
	817 m²

The total surface area of the 11 m X 55 m X 15 m cube equals 3, 190 m².

Therefore,

Blockage Ratio = 817 m² / 3,190 m² X 100 = 25.6 %

Once the blockage ratio is determined, the following guidelines should be utilized as appropriate.

5.5.2.1 Semi-Enclosed Volumes

Partially enclosed volumes consist of partially open buildings, process structures, and areas beneath process equipment, such as fin-fan air coolers and LNG cold boxes. Gas detection should be provided in partially enclosed volumes greater than 1,000 m³ (35,000 ft³) if they have a blockage ratio greater than 0.30.

The detection in a partially enclosed volume should be capable of sensing a spherical gas cloud 5 m (16 ft) in diameter anywhere in the volume.

As indicated earlier, this methodology was initially developed for use in the offshore environment where the semi-enclosed volumes encountered are typically smaller and not as tall as those found onshore. As a result, its application in some onshore applications may require an unusually large and unnecessary number of detectors. The user may want to consider the following alternatives:

- If the volume does not contain a potential release source it may be more effective to monitor the source(s) following the guidance of Section 5.4, Detector Placement for Source Monitoring" or to monitor the perimeter of the volume to detect releases as they migrate into it. If used, the perimeter detection should be arranged as follows:
  - Where the migration hazard consists of heavier-than-air gases from distant sources, it is usually sufficient to provide grade-level detection (18 inches or 0.5 m above grade) along the perimeter of the volume.
  - Where the migration hazard consists of heavier-than-air gases from nearby sources or of neutrally buoyant gases, it is usually sufficient to provide multiple levels of detection at 5 m (16 ft) vertical intervals along the perimeter of the volume. In this case, the height of the detection should extend above the height of any foreseeable gas cloud or plume.

- If the volume contains potential release sources, it may be more effective to apply the source monitoring method (see Section 5.4, "Detector Placement for Source Monitoring"). In this case though, dispersion modeling should be utilized to consider releases oriented in all three planes in order to maximize the potential for detection success. In addition:
  - Obstructed portions of the volume should be monitored to assure that a spherical gas cloud 5 m (16 ft) in diameter cannot exist without contacting a sensor.
  - Detection should be provided in elevated locations if the potential for lighter-than-air or neutrally buoyant materials to collect there exists.

5.5.2.2 Open Volumes

Open volumes, which have obstruction ratios of 0.30 or less, typically do not require combustible gas detection, unless they contain numerous pockets of congestion. The term congestion implies that:

- There are more than three layers or rows of equipment, piping, or structures between the center of the pocket and a reasonably open area (roadway, court yard, access road, etc.); and
- The equipment, piping, and structures in each row are fairly close together, but do not result in the blockage ratio of the pocket exceeding 0.30.

Congested pockets with blockage ratios in excess of 0.30 should be provided detection in accordance with the guidance for partially enclosed volumes.

Where large open volumes encompass congested areas one of the following detector deployment strategies should be utilized:

- Large open volumes with numerous congested areas that have blockage ratios less than 0.3 should be provided with detection capable of sensing a spherical gas cloud 10 m (33 ft) in diameter anywhere in the open volume.
- Large open volumes that contain isolated pockets of congestion with blockage ratios less than 0.3 should be provided with detection capable of sensing a spherical gas cloud of 5 m (16 ft) in diameter anywhere within the pockets.

Detection along the volume's perimeter (using the source monitoring method for detector placement) may be used in lieu of volumetric detection for large open volumes if the following conditions are present:

- The volume has no congested parts with a blockage ratio greater than 0.3.
- The hazard of concern is migration of gas from the volume to other areas.

Perimeter detection would normally be used in onshore facilities, such as gas terminals or oil refineries. Perimeter detection would not normally be used in offshore installations because migration of gas from the installation is not usually a hazard.

5.5.2.3 Applying the Volumetric Approach to Liquid Phase and Heavier-Than-Air Releases

A liquefied gas release can behave very differently from a vapour phase release. First, less mixing occurs with the air. Second, the heat required to evaporate the gas is often more than that available from the surroundings, so released fluids may remain liquid for some time. This is particularly true of LNG releases, butanes in cold climates and most any large volume LPG release where auto-refrigeration is a factor.

The local gas cloud is therefore relatively dense and cold. This gas or vapor tends to slump, especially in low air-movement areas. Throughout the length of the plume, gas at the outer edges is entrained into and carried away by local air movement. LPGs will continue to travel along the ground as they disperse, following the ambient air currents and grade effects. LNG vapors will initially behave much like LPG vapours until they have warmed sufficiently at some point downwind, where the whole plume becomes lighter-than-air and "lifts off".

Where liquid phase or heavier-than-air releases are expected, the following detection should be provided as follows.

- Low-level detectors should be installed in areas that contain inventories of process fluids, which if released to atmosphere, would remain in liquid state for a significant period. This detection should be incorporated into the three dimensional grid utilized in the volume and be installed at no more than 0.5 m (1.5 ft) above the local deck or grade.
- Isolated sources of liquefied gas releases, such as storage area transfer pumps/compressors, storage vessels, cavern well heads, and loading installations (truck, marine and rail) in open areas should be provided with low-level detectors arranged on a 5 m (16 ft) triangular grid around the potential heavy gas release points.

5.6 DETECTOR PLACEMENT FOR ENCLOSURE MONITORING

There are two situations that may warrant the need for gas detection within a building or enclosure. This includes those situations where the structure houses equipment that handles toxic or flammable process materials and those situations where the structures are subject to releases from adjacent plant areas.

5.6.1 Monitoring Enclosures for Toxics

The situation is most often encountered in the form of analyzer buildings into which small bore tubing directs toxic or asphyxiating materials through analyzer equipment. It is common to provide the interior of these structures with one toxic gas detector for each gas of concern, along with an oxygen sensor if asphyxiants are present. These detectors are typically placed at a height of 4-1/2 ft in buildings that have limited or natural ventilation, which puts them within the breathing zone of building occupants.

The toxic gas detectors may be placed in or near the exhaust duct inlets of mechanically ventilated buildings if the following conditions are met:

- The placement of the exhaust pick up is appropriate for the specific gravity of the gas.
  - Within 12 inches of the floor for heavier than-air-gases or vapors
  - Within 12 inches of the ceiling for lighter-than-air gases or vapors
- Mechanical ventilation at 6 or more air changes per hour (ACH) is provided.
- The air velocities in the duct do not exceed the manufacturer's recommendations.

Note: If detectors are to be placed in or near an exhaust duct, verify that the air velocities within the duct will not exceed the manufacturer's recommendation.

The gas detection within these structures is usually configured to send an alarm signal to a continuously manned location and to activate a visible and audible alarm located on the exterior of the building adjacent to its primary entrance. Larger buildings should be provided with alarm devices at each entrance. Where feasible, the gas detection should be interlocked with the sampling streams to provide automatic isolation of the process streams.

There are occasions where process equipment must be located indoors due to weather-related or quality control issues. In this case, the toxic gas and oxygen detectors would be located and configured as described for analyzer buildings. However, the layout and size of the structure may warrant:

- The installation of sensors at multiple locations. In large structures, it may be beneficial to space the detectors out along the normal path of travel through the building. See the guidelines on sensor placement for path of travel and receptor monitoring for further details.
- Evaluation of the ventilation rates and ensuing air currents within the building to determine where best to locate the detectors.
- Providing exterior visible and audible alarms at all entrance ways.

The other common situation is where a building or space (as in a room within an enclosed process structure) is free of internal release sources, but is subject to releases from adjacent process and storage areas. These buildings typical house operator shelters, control rooms, motor control and power distribution centers, or process computer spaces. The hazards associated with this type of location include:

- Sudden Engulfment – In this case a large release engulfs the exterior of the building and may quickly migrate into it via the make up air inlets of its HVAC system or through an open doorway.

- Long-Term Entrainment – In this case a minor release, possibly below hazardous and detectable concentrations, enters the building through its HVAC system over a relatively long time. Depending upon the configuration of the building's HVAC system and the buoyancy of the toxic gas (heavier or lighter-than-air) there is the potential for hazardous concentrations to develop over time.

These buildings are typically provided with toxic gas detection and may be provided with oxygen sensors if engulfment by an asphyxiant is of concern. The following guidelines should be considered in placing and configuring the detection for these structures:

- A detector for each gas of concern should be provided in the make-up air inlet of the HVAC system or outdoors in very close proximity to it. These detectors should be configured to activate an alarm, shutdown the make-up and exhaust fans of the HVAC system and to close the HVAC system's make-up and exhaust dampers. The type of detector utilized should provide the shortest reaction time feasible for the target gas. Ideally, the time needed for the detector, control system and make-up air damper to react should be shorter than that required for toxic gas to enter the building boundary (detector lag time is typically the controlling factor).

  In addition, air inlet detectors should be provided with remote calibration capabilities or suitable access platforms, as they will typically be elevated. This is essential in assuring that the detector remains calibrated and well maintained.

  Verify that the air velocities within the air duct will not exceed the manufacturer's recommendations before placing a detector within a duct.

- A detector for each gas of concern should be located inside the building in close proximity to each entrance door. In doing so, the following guidelines would apply to the placement of the detectors in most buildings:
  – The detectors should be placed at a height of approximately 4-1/2 ft to place them within the breathing zone of building occupants.
  – The detectors need to be located within the normally conditioned areas of the building. They should not be located in vestibules, air locks, closets, or small offices that may be regularly closed up. The intent is to keep them in spaces that are subject to normal air mixing and movement.
  – These detectors should be configured to activate an alarm, shutdown the make up and exhaust fans of the HVAC system, and to close the HVAC system's make up and exhaust dampers.

The alarms activated by the gas detection should send a signal to a continuously-manned location (this may not be feasible if the building is the normal alarm receipt point for the plant) and should include visible and audible alarm devices on the exterior of the building at each entranceway and within the structure. The devices within the structure should be configured using the guidance for fire alarm devices, as provided in NFPA 72, *National Fire Alarm Code®*. One important note is that a means of silencing the alarms must be provided within the building, as conditions may dictate that personnel remain within the building until the situation outside the building has been brought under control.

5.6.2 Monitoring Enclosures for Flammables

5.6.2.1 Enclosures Containing Release Sources

The situation is most often encountered in the process industries in the form of analyzer buildings into which small bore tubing directs process materials through analyzer equipment. It is common to provide the interior of these structures with at least one combustible gas detector. The location of the detector(s) should be based on the vapor density of the gases present, with detectors being placed within 12 inches of the floor for heavier-than-air gases and within 12 inches of the ceiling for lighter-than air gases.

The combustible gas detectors may be located in the exhaust duct inlet of mechanically ventilated buildings if the following conditions are met:

- The placement of the exhaust pick up is appropriate for the specific gravity of the gas.
  - Within 12 inches of the floor for heavier than-air-gases or vapors
  - Within 12 inches of the ceiling for lighter-than-air gases or vapors
- Mechanical ventilation at 6 or more air changes per hour (ACH) is provided.
- The air velocities in the duct do not exceed the manufacturer's recommendations.

The gas detection within these structures is usually configured to send an alarm signal to a continuously-manned location and to activate a visible and audible alarm located on the exterior of the building adjacent to its primary entrance. Larger buildings should be provided with alarm devices at each entrance. Where feasible, the gas detection should be interlocked with the sampling streams to provide automatic isolation of the streams upon activation of an alarm.

There are occasions when process equipment must be located indoors due to weather-related or quality control issues. Combustible gas detection should be provided in these buildings using one of the deployment options discussed below. This detection should be configured to:

- Send an alarm signal to a continuously manned location
- Activate visible and audible alarm devices on the exterior of the building at each entranceway and within the structure. The devices within the

structure should be configured using the guidance for fire alarm devices, as provided in NFPA 72, *National Fire Alarm Code®*.

- Include a local means of silencing the alarms.

In some cases, it may also be advantageous for the detection system to initiate automatic shutdown and/or isolation of the potential release sources within the structure. However, the risk of spurious shutdowns due to false or nuisance alarms should be closely evaluated to assure that the automatic controls will not increase the risks to the facility. In many cases, remote operated shutdown and isolation systems and/or remotely located isolation means will suffice.

The first deployment option would be to lay out the gas detectors based on the specific gravity of the gases present and the air movement within the structure as discussed below. As an alternative, the volumetric approach could also be utilized.

Heavier-than-Air Gases and Vapors. If the structure is naturally ventilated, a minimum of four gas detectors should be provided. These detectors should be arranged in a square grid with a maximum diagonal measurement of approximately 13 ft (4 m) for a single high potential release source (or closely spaced multiple source). Where more than one source is present, the detectors should be arranged in a square or rectangular pattern that is supplemented by additional detectors so that the maximum distance between any two detectors is 13 ft (4 m).

In mechanically-ventilated structures (6 ACH or greater), the gas detectors should be laid out based on expected path of vapor travel. Once the structure has been substantially completed, the air movement through it should be evaluated using very fine powders (commercially available for this purpose), air velocity meters, or fine strips of tissue paper to identify areas along the floor where air currents are normally present or absent. Once this is accomplished:

- Gas detection should be placed in dead air spaces that encompass potential release sources.
- Gas detectors should be placed in those locations where the air streams converge and approach the inlets of the exhaust ventilation system. The detectors should not be placed in or too close to the duct work, unless it is verified that the air velocities will not exceed those recommended by the manufacturer. It may be necessary to use more than one detector at each inlet to assure an adequate probability of detection success.

Additional detectors should be placed in pits, trenches or other low areas.

Lighter-than Air Gases and Vapors. If the structure is naturally ventilated, gas detectors should be located within 12 inches of the ceiling. Where the ceiling is pitched, the detectors should be located along a line approximately 18 inches

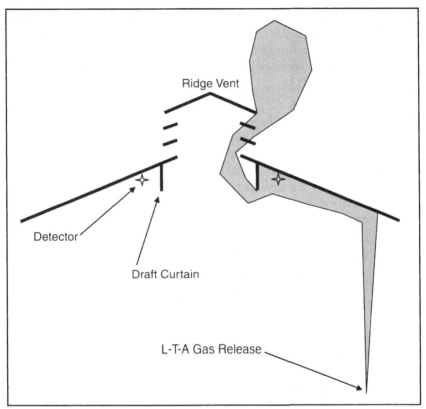

Figure 5.3 — Detector Placement Relative to Draft Curtain Placement for Lighter-than-Air Gases and Vapors

(0.5 m) on each side of the ridge with a maximum distance along the ridge line of approximately13 ft (4 m) between detectors. Where the building has been provided with a ridge vent, consideration should be given to providing a solid draft curtain between the ridge vent and each line of detectors, as shown in the following figure. The draft curtain should have sufficient depth to maximize the potential of a release accumulating and contacting a detector, but should not inhibit the overall air flow of the structure.

If the ceiling is flat, the detection should be arranged in a square grid centered over the source with a maximum diagonal measurement of approximately13 ft (4 m) for a single high potential release source. Where more than one source is present, the detectors should be arranged in a square or rectangular pattern that is centered over the sources and supplemented by additional detectors so that the maximum distance between any two detectors is 13 ft (4 m).

In mechanically ventilated structures (6 ACH or greater) the gas detectors should be laid out based on expected path of vapor travel. Once the structure has been substantially completed, the air movement through it should be evaluated using very fine powders (commercially available for this purpose), air velocity meters, or fine strips of tissue paper to identify areas along the ceiling where air currents are normally present or absent. Once this is accomplished:

- Gas detection should be placed in dead air spaces that encompass potential release sources.
- Gas detectors should be placed in those locations where the air streams converge and approach the inlets of the exhaust ventilation system. In doing so, the detectors should not be placed in or too close to the duct work, unless it is verified that the air velocities will not exceed those recommended by the manufacturer. It may be necessary to use more than one detector at each inlet to assure an adequate probability of detection success.

Where the ceiling contains beam pockets, draft curtains, or other obstructions that would prevent the free movement of the vapors along the ceiling additional detectors should be located within these areas.

Neutrally Buoyant Gases and Vapors. Due to the unpredictability of vapors in this application, it is suggested that the volumetric approach to detector placement be utilized for neutrally buoyant vapors and gases.

5.6.3 Enclosures Exposed to External Release Sources Only

It is often necessary to locate buildings and other structures that house potential ignition sources in process and storage areas. These buildings are usually pressurized to prevent flammable vapors from contacting the ignition sources, and often house motor control centers, distributed control equipment, operator shelters, and control rooms.

Combustible gas detection should be considered for all structures housing ordinary hazard electrical equipment, fired equipment (hot water heaters, boilers, etc.) and other potential ignition sources if they are located in an electrically classified area or within the plume of a credible release scenario. This combustible gas detection is intended to protect the structure from:

- Sudden Engulfment – In this case a large release engulfs the exterior of the building and quickly migrates into it via the make-up air inlets of the HVAC/pressurization system or through open doorways.
- Long-Term Entrainment – In this case, a minor release that may be below hazardous and detectable concentrations enters the building (usually through its HVAC/pressurization system) over a relatively long time. Depending upon the buoyancy of the gases involved (heavier or lighter-than-

air) and the configuration of the building's HVAC system, there is the potential for flammable concentrations of gas to accumulate over time.

The following guidelines should be considered in placing and configuring the detection for these structures:

- A combustible gas detector should be provided in the make-up air inlet of the HVAC/pressurization system or outdoors in very close proximity to it. This detector should be of a fast response technology, such as IR, and configured to:
  - Activate an alarm, shutdown the make-up and exhaust fans of the HVAC system, and close the HVAC system's make-up and exhaust dampers. The total response time of the detector, control system, and make-up air damper should be shorter than the total time needed for the gas to reach the area boundary after passing the detector.
  - Be remotely calibrated or provided with suitable access platforms, as these air intakes are typically elevated. This is essential in assuring that the detector remains calibrated and well maintained.

  Verify that the air velocities within the air duct will not exceed the manufacturer's recommendations before placing a detector within a duct.

- A gas detector should be located inside the building in close proximity to each entrance door. In doing so, the following guidelines would apply to the placement of the detectors in most buildings:
  - The detectors should be placed at a height of approximately 12 inches (0.3 m) above the floor where heavier-than-air gases are of concern or approximately 12 inches below the ceiling if lighter-than-air gases are of concern.
  - The detectors need to be located within the normally conditioned areas of the building. They should not be located in vestibules, air locks, closets, or small offices that may be regularly closed up. The intent is to keep them in spaces that are subject to normal air mixing and movement.
  - These detectors should be configured to activate an alarm, shut down the make-up and exhaust fans of the HVAC system and to close the HVAC system's make up and exhaust dampers.

The alarms activated by the gas detection should:

- Send an alarm signal to a continuously-manned location (this may not be feasible if the building is the normal alarm receipt point for the plant).
- Include visible and audible alarm devices on the exterior of the building at each entranceway and within the structure. The devices within the structure should be configured using the guidance for fire alarm devices, as provided in NFPA 72, *National Fire Alarm Code®*.

- A local means of silencing the alarms must be provided within the building, because conditions may dictate that personnel remain within the building until the situation outside has been brought under control.
- The alarms also may initiate the automatic shutdown of fired equipment, ordinary hazard electrical equipment, and other ignition sources where the risk of spurious or nuisance shutdowns from false alarms can be tolerated. In some cases, this may be required by the building's electrical area classification scheme.

5.7 DETECTOR PLACEMENT FOR PATH OF TRAVEL AND TARGET RECEPTOR MONITORING

Detector placement for path of travel and target receptor monitoring is typically used in toxic gas applications and may be used outdoors and inside very large enclosures.

This method of detector placement may also be used for flammable gas applications where other placement methods may not adequately protect emergency egress paths and marshalling points. It may also be useful for protecting the entry point and parking area utilized by personnel who must serve remotely located facilities.

In applying this placement methodology, the objective is to provide detection where personnel are expected to travel or congregate on a regular basis. In general, detectors are placed as follows:

- Along normal paths of travel within the unit/area, particularly in those locations where personnel may not be able to observe the area as they approach potential release sources.
- Near normal points of entry into the unit/area. In this case, detection would be placed along the path of travel from entry gates, operator shelters, or other centralized locations from which most people are expected to enter the unit/area.
- In confined spaces, partially enclosed areas, and enclosed buildings that house equipment in toxic or asphyxiating gas service.
- In below grade locations that are normally accessed by personnel, particularly if heavier-than-air and neutrally buoyant gases are present.
- Between equipment in toxic service and primary evacuation marshalling points, if a credible release may affect the marshalling point.
- Between the potential release sources and the uncontrolled areas such as parking lots, plant road ways and buildings unrelated to the process/storage area.

Grade level toxic detection should generally be installed at a height of 18 inches for heavier than air gases, as released gases are more likely to reach these locations sooner and to accumulate at higher concentrations. Toxic detection for gases that are at or close to neutral buoyancy should be installed at a height of approx. 4-1/2 ft to place them close to the breathing zone of personnel.

In addition to the grade level detection outlined above, protection should be provided for elevated structures. This additional protection should consist of detectors placed immediately adjacent to the stair landings and ladders that are normally used to access each deck. At least one detector should be placed approximately 18 inches above the deck with a second detector being placed approximately 4-1/2 ft above the deck if the gases are expected to be neutrally buoyant. The intent of placing the detectors in this location is to alert personnel of releases that may affect their breathing zone as their head first crests the deck and before they have had a chance to observe the conditions on the deck.

In remotely located or unmanned locations, toxic gas and/or flammable gas detection should be provided along the normal path of approach to the site or the point where personnel will normally disembark from their vehicles.

It is typically not practical to provide fixed toxic gas detection for lighter-than-air gases, unless some type of roof or solid deck structure is present above the source(s). Otherwise the available footprint in which a detector must be located is likely to be too small in cross sectional areas and unpredictable. The one exception to this would be in the case of refrigerated or cryogenic materials, which may initially behave like a heavier-than-air or neutrally buoyant material. In this case, dispersion modeling should be utilized.

5.8 DETECTOR PLACEMENT FOR PERIMETER MONITORING

Perimeter protection is usually not necessary where only lighter-than-air hazards are being handled as these materials are not likely to travel far from their source in the horizontal plane.

Perimeter monitoring is sometimes utilized where heavier-than-air or neutrally buoyant hazards are present, to notify personnel that a credible release scenario is affecting an area that is regularly occupied by:

- Strong ignition sources that are outside of the owner's control. This may include facilities utilizing fired equipment, fixed internal combustion engines, welding operations, and those areas frequented by motor vehicles, shipping, or rail traffic.
- Occupied by persons that are not under the owner's control or not expected to be familiar with the hazards posed by potential releases or the emergency procedures for protecting themselves from them.

Perimeter protection is sometimes mandated by regulatory officials or implemented by facility management interested in providing increased protection for the public. In many cases however, it would be more effective, from the perspectives of reliability and cost, to provide the localized detection discussed elsewhere in this chapter or to increase the density of detectors in those areas because:

- Localized detection can effectively detect smaller releases than perimeter protection.
- The number of variables that come into play increases exponentially as the release moves away from the source, making it increasingly difficult to predict its behavior and assure a high probability of detection.
- The lack of utilities and plant infrastructures (cable trays, conduits, equipment supports, etc.) along most fence lines can significantly drive up the cost of these systems. This is particularly true where laser detection is not applicable and point type detectors must be used.

There is one case in which the use of perimeter protection may be effective enough to warrant its installation. This is where heavier-than-air flammable vapors are being handled and:

- The process or storage is relatively close to the perimeter and dispersion modeling indicates that small, but high probability releases may cross the fence line; and
- The sources of such releases are outside the coverage of local detectors and/or very close to the limits of the local detection coverage.

Perimeter detection for heavier-than-air flammable vapors/gases is generally laid out in accordance with the general guidance and recommendations for source containment provided earlier in this chapter. Additional considerations include:

- Alarm set points should be as low as practical without causing nuisance alarms.
- It is typically more effective to utilize IR or laser beam detection in this application, with point detectors being used to cover gaps in the beams and areas that are not conducive to the use of beam protection.
- This detection is typically located in un-maintained areas, which may necessitate the detectors/beams being located at higher elevations than normal in order to accommodate snow build up, vegetation and wildlife.
- The height of IR or laser beam sets must be selected based not only on the density of the gas, but also on the highest point of grade along the beam, in order to prevent false detections and blocked beam alarms from wildlife, grass, and vegetation.
- The higher the detector/beam must be located above grade, the lower its alarm set point will need to be set in order to remain effective

Where feasible, it may be more effective to locate the perimeter protection somewhere between the release sources and the property line, as opposed to placing them directly on the property line. This should allow for the detection of smaller releases and provide more time for responding to them.

5.9 DETECTOR SET POINTS AND MONITORING

All gas detection should be tied into an alarm system that:

- Provides visible and audible alarm signals that are distinctive from other alarms in the plant. It is essential that a perceived alarm convey a single message to plant personnel, with regards to the cause of the alarm and the actions that they are to take.
- Transmit an alarm signal to a continuously-manned alarm monitoring point within the plant such as a control room, guard house, or emergency operations center.
- Transmit an alarm signal to the local operator shelter or control room, if applicable.

Ideally, gas detectors should be continuously monitored by a distributed control system (DCS) or manufacturer-approved panels that allow for instantaneous data acquisition and trending capabilities over their entire calibrated range.

Note: Most detectors come with internal alarm relays in addition to the 4-20 mA outputs. The disadvantages of using the alarm relays to communicate alarms are that the user is provided with no:

- *History upon which they can select alarm set points or evaluate incident histories (The lack of an alarm during a release does not necessarily indicate a detection failure, it more than likely means that gas concentrations at its location were below its alarm set points); and*
- *Opportunity to use time-delayed or rate-of-change alarm configurations to better monitor for low-level gas concentrations (avoiding false alarms) or abnormal changes that may be indicative of an accidental release.*

Alarm set points should be set as close to ambient conditions as possible without causing nuisance alarms, so that the probability of detection success and personnel protection may be maximized.

The alarm set point for installations involving unfamiliar conditions (new type of detectors, operations, or process equipment) should be established as follows:

- Initially set the alarm concentration at a regulated level, such as the OSHA PEL or 50% of the LEL and retain it for a minimum of 30 days.
- Review the trend data to identify the peak and average concentrations of the material present.
- Incrementally adjust the alarm set point downwards until the minimum setting recommended by the manufacturer is reached or nuisance alarms become an issue. If nuisance alarms are experienced, the set point should be adjusted upwards to the last setting at which they did not occur.

This may be accomplished by retaining the high-high alarm at the last setting that provided no nuisance alarms and then resetting high-alarm downward to the next incremental setting. If desired, the high alarm could be programmed as un-annunciated and used for data tracking purposes only.

The alarm set point of outdoor detectors will typically need to be set higher above ambient conditions (i.e., greater safety factor against nuisance alarms) than indoor detectors, as they may be more prone to spurious and temporary electronics deviations (that could result in nuisance alarms where the set points are at or near the limits of the device's detection range or span), because they are exposed to less stable atmospheric conditions such as rapid changes in temperature, humidity, and solar radiation.

Where periodic or short-term excursions are normally experienced, the use of two alarm set points may be utilized to avoid nuisance alarms. In this case, the DCS or control panel could be programmed using one of the following alarm configurations:

- In the first case, two alarm set points would be utilized. Alarms would be annunciated immediately at the high-level threshold and at the low low-level threshold only if the gas concentration lasts longer than normally expected excursions.
- In the second case, a single alarm set point and trend analysis would be utilized. Alarms would be immediately annunciated if the alarm threshold level were exceeded or if the concentration of gases increased faster than expected based on the trend analysis.

There is no need for consistent alarm point settings to be utilized throughout an entire facility. It is preferable that areas with high ambient concentrations of target gas or cross-sensitivity gases be handled as special cases, instead of applying facility-wide set point adjustments. Doing so only serves to reduce the level of protection provided outside of these special areas.

In no case should the set point of a toxic detection system be higher than the TWA threshold defined by local policies or legislation.

6

Overall System Management — Commissioning, Testing, and Maintenance

6.1 SUMMARY

The previous chapters have reviewed the need for detection, the options for detection, and items to consider when designing your system. This chapter will review what it takes to commission and maintain a gas detection system. This will cover initial acceptance testing, training requirements, and on-going system maintenance and testing.

6.2 TRAINING

If possible, one should determine training requirements early in a project. Design, installation and testing are great training opportunities. So, if the requirements are known, training can be integrated into these phases of the project.

The types of training that may be required include:

1) System design, layout, and detection philosophy.
2) Detector troubleshooting and maintenance.
3) Control/monitoring system troubleshooting and maintenance.
4) Response:
   a) How will the Emergency Response Teams (ERT) get the information they need in order to effectively respond to the event in a safe manner? Educating the responders on how to obtain the correct information is critical. They will need to know what the appropriate Personal Protection Equipment (PPE) will be and what Handheld Gas detection units to respond with.
   b) What evacuations will occur automatically/manually?
5) Alarm Silence procedures.
6) Resetting the system.
7) Post event follow-up that is required before normal operations can resume.

6.3 DOCUMENTATION

Good documentation practices are essential. The facility should have an established standard for documentation that addresses both format and content. All documentation should be easily accessible in one location. Establishing documentation can be burdensome; however, the paybacks can be immense (troubleshooting a system, trying to explain to a new technician how the system is designed and functions, etc.) Good documentation is a critical step towards establishing credibility with the local authority-having jurisdiction (AHJ).

Records should be kept of:

- Personnel training
- All calibration, maintenance and troubleshooting activities
- System configurations and system capacities
- System drawings
- And all those unique features of one's system (tribal knowledge)

All documentation should be current and up to date. Documentation should be reviewed periodically to ensure that it is still current. Bad documentation can be worse than no documentation since it can lead to bad decisions that could result in injuries and damages.

6.4 MAINTENANCE

One should establish a maintenance schedule that is feasible for the organization. Planning for all calibrations to occur during an annual shutdown may work for one organization but not another. In some cases, it is best to spread the workload out over the entire year. This all depends on the resources available and the number of detectors that have to be maintained. Regardless of how one's schedule is established, the system must be maintained in accordance to the manufacturer's recommendations. This is critical, because failure to do so results in not only increased risk of failure and false alarms, but also increased liability if something goes wrong (i.e., "Why did you fail to keep the system calibrated?")

Some businesses will choose to contract the maintenance out to a qualified service organization. This alternative can sometimes be cheaper than gaining the required expertise in-house.

The system should be designed and installed with maintenance in mind. Placing detectors in locations that are not easily accessible is never a good idea. It can lead to maintenance not being done properly or on schedule.

Thirty-Minute Guideline

The International Fire Code requires that toxics be monitored continuously. It then describes continuous as "every 30 minutes." So if maintenance can be completed in less than 30 minutes, the code does not require that equipment or gas sources be idled/shut off.

Part of a good maintenance plan will also include ready access to critical spares. One should evaluate potential downtime for the various failure modes. If unscheduled downtime is devastating to the business, one will want to ensure that the parts and personnel are available to repair the system within the 30-minute window.

6.5 ESTABLISH A GOOD RELATIONSHIP WITH THE LOCAL AUTHORITY-HAVING JURISDICTION (AHJ)

When managing the system, make sure that the AHJ realizes all of the efforts that have been made in establishing it. It takes time, but if one works to instill confidence with the AHJ through demonstrated competence, it can make one's life much easier in the long run. Some examples of these are:

1) If deviations from the letter of the code are requested via an alternative means and methods, the AHJ is more likely to approve these requests if they have the confidence that the item can be managed in the manner that has been proposed.
2) Testing: The AHJ has the right to witness all testing of the system. Scheduling the AHJ's time can sometimes be an issue on a fast-track project. On the other hand, if one's competence has previously been demonstrated and the testing documentation is in order, the AHJ might waive his right to witness the testing since they have confidence in the facility's abilities and the overall program.

6.6 CHANGE MANAGEMENT

Like all systems, change is inevitable. Process and system changes will require the addition and/or deletion of detection components. A common mistake is to focus on these changes as isolated items. The risk of this approach is that over the life of a facility, all of the "isolated" changes will have completely changed the overall system design. If drawings and documentation have not been kept updated, when an issue occurs, troubleshooting becomes much harder and more time consuming and the risk of production impacts goes up.

In a different scenario, the system has always been treated as a whole unit. All changes were evaluated on their impact to the overall system, including training, maintenance, and documentation. Now when there is an issue that requires troubleshooting, the documentation can be used as an aid to can greatly reduce the amount of time to get the systems back on line. Or, if asked about the impact of a proposed major renovation, one hasyou have all of the information available to make informed decisions about the scope of work that will be required for the gas detection system.

There are several methods to keep track of the system. It could be as simple as a spreadsheet, or as elaborate as a database with drawings, test sheets, and system programming.

Basic features that should be included are:

- Point IDs: Identifiers for all devices in the system (Detectors, Horns, Strobes, Shutdown relays) and their location.
- Capacity: Determining the installed and spare capacity.
- Dates: Last test date andTest Date upcoming maintenance.
- Response: What happens when a gas release is detected?

Examples:

HPM System Testsheet

MIM436

Test Date: 12/6/2005

Testing Witnessed by:

Test Command:

Program Verified: _____

Vertex sheet updated: _____

Graphics Verified: _____

☐ *hierarchy* ☐ *faceplates*

Fab Ops:

SubFab Ops:

MDA Ops:

ENG USE ONLY: Typical: Implant   Fab Level   Sector H   ☐ *Run Seq/Name* ☐ *CFC* ☐ *Graphics* ☐ *Data Archive*

☐ *Alarm* 11150108   D1D-VRTSW-01-08 - MIM436 GAS BOX SOURCE - FAB LEVEL BAY 200 - BF3   VRT_BF3_Gas ☐
   ☐ 10570805   D1D - MIM436 IMPLANTER #5 TOOL S/D SIGNAL   Tool_S/D ☐
   ☐   ALARM LEVEL/MESSAGE CLASS: Alarm
   ☐   HORN/STROBES   Bay: Yes HS_Bay200   Chase: Yes HS_Chase200N   Other: No

SID Compliance? Yes / No _____	*(Signature)*	*Gas Type* BF3	Alarm	250.0
Flow Tested? Yes / No _____	*(Signature)*	*Unit:* ppb	Warning	150.0
Labeling? Yes / No _____	*(Signature)*		Max	1500.0
Specific Notes:				

☐ *Alarm* 11150116   D1D-VRTSW-01-16 - MIM436 GAS BOX SOURCE - FAB LEVEL BAY 200 - AsH3 PH3   VRT_AsH3_Gas ☐
   ☐ 10570805   D1D - MIM436 IMPLANTER #5 TOOL S/D SIGNAL   Tool_S/D ☐
   ☐   ALARM LEVEL/MESSAGE CLASS: Alarm
   ☐   HORN/STROBES   Bay: Yes HS_Bay200   Chase: Yes HS_Chase200N   Other: No

SID Compliance? Yes / No _____	*(Signature)*	*Gas Type* AsH3	Alarm	50.0
Flow Tested? Yes / No _____	*(Signature)*	*Unit:* ppb	Warning	25.0
Labeling? Yes / No _____	*(Signature)*		Max	500.0
Specific Notes:				

ISSUES/VIOLATIONS REPORT:

_____   *Resolution:* _____

HPM System Testsheet

VMB Cl2-i4J/o/p-B/8.5e

Test Date _____ 1/26/2001

Testing Witnessed by:

Test Command:

Program Verified: _____

MDA/CM4 sheet updated: _____

Fab Ops:

SubFab Ops:

MDA Ops:

IRC-3

☐ 062674   F20-MDAG-07-14 ABS GAS CABINET Cl2-i4J-J FEEDING VMB'S B-8.5e & tbd - Cl2   ◉
  ☐ 061603   F20 - GAS PAD AREA HORN   ◉
  ☐ 061604   F20 - GAS PAD AREA STROBE   ◉
  ☐ 061605   F20 - GAS PAD LOADING DOCK AREA HORN   ◉
  ☐ 061606   F20 - GAS PAD LOADING DOCK AREA STROBE   ◉
  ☐ 061607   F20 - GAS PAD STAGING RM HORN   ◉
  ☐ 061608   F20 - GAS PAD STAGING RM STROBE   ◉
  ☐ 061609   F20 - GAS PAD FLAM RM HORN   ◉
  ☐ 061610   F20 - GAS PAD FLAM RM STROBE   ◉
  ☐ 063044   F20 - ABS GAS CABINET Cl2-i4I SHUT DOWN   ◉
  ☐ 063045   F20 - ABS GAS CABINET Cl2-i4J SHUT DOWN   ◉
  ☐ 065673   F20 - HIGH LEVEL ALARM TO SECURITY - CALL A CODE   ◉

SID Compliance? Yes / No _____   *(Signature)*

Flow Tested? Yes / No _____   *(Signature)* ◉

Pt. Online/Labeled? Yes / No _____   *(Signature)* ◉

Specific Notes (SubFab)

SAMPLING DEVICE (if applicable)	F20-MDAG-07-14	
	Gas A	Gas B
Low	0.25 ppm Cl2	
High	0.5 ppm Cl2	

Index

Printed in the United States
By Bookmasters